International Oil Company Financial Management In Nontechnical Language

International Oil Company Financial Management In Nontechnical Language

by James Bush *and* Daniel Johnston

PennWell
1421 South Sheridan Road
P. O. Box 1260
Tulsa, Oklahoma 74101

Library of Congress Cataloging-in-Publication Data

Bush, James W.
Johnston, Daniel

Printed in the United States of America
1 2 3 4 5 02 01 00 99 98

Dedications

Dedicated forever with love to my wife, Lorraine,
our children and grandchildren, and
my associates in the oil and gas industry.

To my wife Jill and our children,
Erik, Lane, Jill Danielle, Julianna, David, Sheridan,
our foster children, Lucero and Margarita,
and to our parents.

Contents

Figures

Tables

Acknowledgement

The illustrations in this book were created by Daniel's twin brother, David Johnston, a management consultant and electronic engineering graduate of the University of Rochester, New York.

Introduction

This book is written for those interested in learning about the function of financial management of petroleum companies. We have attempted to provide a glimpse of the key aspects, issues and concepts that dominate this scene. Furthermore, we have tried to provide a source of reference for these concepts as well as the terminology associated with this subject.

Financial Management for the Petroleum Industry

1

The financial concepts in this text go beyond the traditional finance departments that in the past have been solely responsible for collecting, reporting, interpreting, evaluating and making the financial decisions of the firm. The primary purpose is to provide a guide to the theories, quantitative methodologies, and step-by-step application of these concepts in the area of financial management in a nontechnical format.

Individuals can utilize this text as a reference or as a guide to assist in expanding their financial knowledge base. It should be useful to anyone interested in learning more about the functions of finance. Utilizing current computer systems allows easy access to financial information throughout the company. Such access of real time information is a boon to any company manager who must evaluate the firm's financial status for decision-making purposes.

This chapter provides an overview and history of the development of financial procedures and how they affect the petroleum industry. In addition, the interdependencies of price, supply and demand, and their critical implications are highlighted.

The Importance of Understanding International Finance

An understanding of international finance is crucial to not only the large oil companies, numerous subsidiaries, and joint ventures, but also to the small oil companies engaged in exporting, importing, or other international operations. Of the 43,300 U.S. firms that export, 78% have less than 100 employees.

International finance is even important to companies that have no intention of engaging in international oil business. These companies must recognize how their foreign competitors will be affected by movements in price, supply and demand of crude, exchange rates, foreign interest rates, labor costs, and inflation. Such economic characteristics can affect the foreign competitors' cost of production and market pricing policies.

Companies must also recognize how domestic competitors who obtain foreign crude or foreign financing will be affected by economic conditions in those countries. If domestic competitors are able to reduce their costs by capitalizing on opportunities in international markets, they may be able to reduce their prices without reducing their profit margin. This could allow them to increase market share at the expense of the purely domestic companies.

Most chief executive officers (CEOs) and managers of today's multinational oil companies (MNOCs) recognize the importance of international business. Many of them have been heavily involved in foreign projects and therefore have a more global view than their predecessors. New CEOs will likely be more willing to transfer operations around the globe, which may cause more layoffs in particular countries. The new CEOs are more focused on gaining a competitive edge, which requires a global perspective.

PETROLEUM INDUSTRY PIONEERS IN GLOBALIZATION CONCEPTS

As the petroleum industry matured, it developed by necessity an awareness of opportunities in foreign markets. Over time, supply and demand became displaced geographically and the need followed to export crude to or import products from a foreign firm. Therefore, in the early part of the 20th Century the industry became global.

The Middle East accounts for the lion's share of petroleum reserves and production, while the West consumes oil in great disproportion to its contribution to supply. The reason for the West's large appetite for oil is, of course, the United States, the world's single biggest oil market.

	Reserves		Production		Consumption	
Region	Billion BBLS	Percent	MMBPD	Percent	MMBPD	Percent
Eastern Hemisphere						
Middle East	659.5	64.9%	20.1	29.7%	3.9	5.7%
All Other	191.9	18.8%	28.1	41.6%	39.6	58.4%
Total	851.4	83.7%	48.2	71.3%	43.5	64.1%
Western Hemisphere						
USA	29.6	2.9%	8.3	12.3%	17.0	24.9%
All Other	135.9	13.4%	11.1	16.4%	7.5	11.0%
Total	165.5	16.3%	19.4	28.7%	24.5	35.9%
Total World	**1,016.9**	**100.0%**	**67.5**	**100.0%**	**68.0**	**100.0%**
OPEC	**778.2**	**76.5%**	**27.6**	**40.9%**	**6.9**	**10.1%**
All Other	**238.7**	**23.5%**	**39.9**	**59.1%**	**61.0**	**89.9%**

Table 1–1 Worldwide Distribution of Oil Reserves, Production and Consumption
Source: BP Statistical Review of World Energy-1996

In every measure of the world's petroleum supply, the Middle East prevails. At the beginning of 1995, the region had 660 billion barrels of estimated reserves out of the worldwide total of roughly 1 trillion barrels. Of the Middle East reserve total, countries bordering the Arabian Gulf accounted for a full 98%. The Middle East's 1995 production of 20.05 million barrels per day, (MMBPD) of crude oil—92% of it from Arabian Gulf countries is approximately one-third of the world's total of 67.5 MMBPD. The region's importance to worldwide supply is further illustrated by the concentration of so-called supergiant fields within it. As a market for oil products, however, the Middle East barely rates. The region's consumption in 1995 averaged only 3.9 million B/D versus the world's total of 67.9 MMBPD.

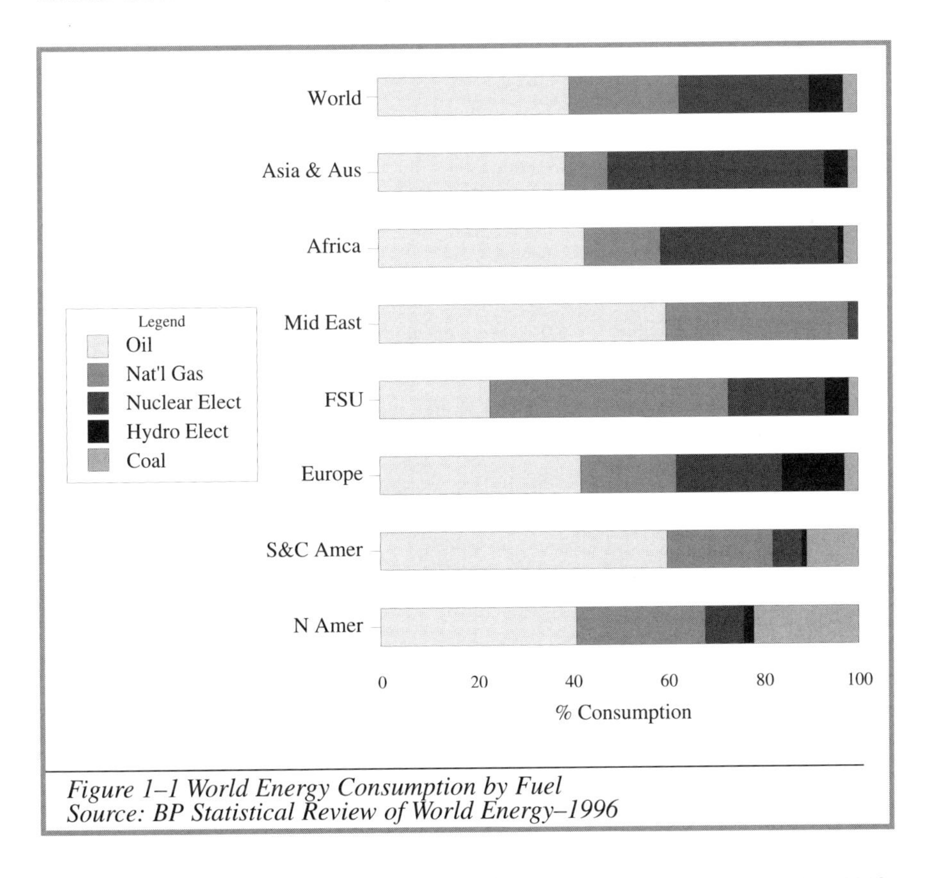

Figure 1–1 World Energy Consumption by Fuel
Source: BP Statistical Review of World Energy–1996

The demand for oil is forecasted to grow from 68 MMBPD in 1995 to 90+ MMBPD in 2010. Most of the growth in demand will be in Organization for Economic Cooperation and Development (OECD) and Emerging Market Economies (EME) member nations. It is expected that the demand growth of 21 MMBPD will be produced by OPEC member countries.

The petroleum industry has invested millions of dollars in transportation, storage, and refinery upgrades to accommodate supply, demand, and seasonal variations. To many the "international oil industry" has meant mostly upstream exploration. Now more development or redevelopment/rehabilitation and projects farther downstream are available in places like Indonesia, China, former Soviet Union/Commonwealth of Independent States (FSU/CIS), and the Eastern Bloc.

An example of a unique international joint venture is the recent announcement that the Amoco and Tractebel groups are joining with the Hashemite Kingdom of Jordan. The venture's objectives are to develop, construct, own, and operate a pipeline system within Jordan to transport, distribute, and market natural gas. The venture is planning to move natural gas from the development of new-found reserves in the Nile Delta to consumers, which include Jordan's fertilizer industry and other major consumers, such as electricity-generating power stations.

FINANCIAL MANAGEMENT OBJECTIVES

The commonly accepted objective of an MNOC is to "maximize shareholder/owner's wealth" which sometimes seems a little trite. But developing an objective is necessary as it requires management processes and decisions contributing to its accomplishment.

Owner/Shareholder's value is represented by the market price of the company's common stock or the value of the firm. The value is a reflection of the firm's financial policies of investment, financing, and distributions to the owners. Therefore, management must develop strategies and make decisions that optimize long-term growth and short-term profits. Any proposed corporate strategy should consider both potential earnings and risks inherent in the strategy. Strategies should be implemented only when the benefits to be derived outweigh costs and risks.

It has often been argued that managers of a firm sometimes make decisions that conflict with the firm's objective to maximize shareholder wealth. For example, a decision to establish a subsidiary in one location versus another may be based on the location's personal appeal to the manager rather than on its potential benefits to shareholder/owners. In order to receive more responsibility and compensation, decisions to expand may be driven by the desires of managers to make their respective divisions grow at the expense of other divisions. If a firm were composed of only one owner, who was also the sole manager, a conflict of goals would not occur. However, for corporations with numerous shareholders who differ from their various managers, a conflict of goals can easily exist. This conflict is often referred to as the agency problem.

Creative financial management is required to survive the industry's challenges. These challenges include competition for available funds, trends toward more hostile operating environments, volatile prices, and governmental control. With U.S. domestic oil and gas boom days over and higher levels of risk and uncertainty, it is critical that an oil and gas company understand the importance of the role of finance and its contributions to the success of the firm.

Financial Strategies

In order to achieve the firm's financial objective of maximizing shareholder/owner's wealth, it is necessary to develop strategies. The strategies will include the following:

- Providing dependable sources of funds at the lowest cost
- Implementing the best financing mix or capital structure for the firm
- Balancing the value of distribution of funds to investors against the opportunity cost of the retained earnings used as a means of equity financing
- Managing working capital in order to maximize profitability relative to the funds tied up in current assets
- Choosing the appropriate sources of funds
- Utilizing sharing methods to obtain funds for exploration and development in advance of production without the traditional debt restrictions

The Economic Environment and Its Impact upon Financial Strategies

Strategies of financial management are affected by the uniqueness of the economic environment, life cycle stage of the industry or enterprise, and the expectations of the owners. Strong inflationary economies have pushed interest rates to unprecedented heights and the resulting high cost of capital has led to profound changes in corporate financial policies and practices. On the other hand, recessionary economies have decreased the demand for capital resulting in lower interest rates and changing the financial policies and practices accordingly.

In the early 1900s, when industrialization was sweeping the world, companies were faced with the critical problem of obtaining capital for expansion. Capital markets were relatively primitive making the transfer of funds from individual savers to businesses difficult.

In the oil and gas industry, a unique basis has been used frequently for raising cash. The high risk, capital intensive nature, and "make-it or break-it" attitude has generated interesting partner relationships between those who have cash and those who have "know-how." However, as the industry matured, companies became larger and integrated the various segments of the industry into megacorporations.

Financing Intermediaries

Most capital transfers occur through specialized financial institutions which serve as intermediaries between sources and users of capital. Specialized financial intermediaries include:

- Commercial banks are the department stores of finance, serving a wide variety of savers and those with needs for funds.
- Mutual savings banks accept savings primarily from individuals, and lend mainly on a long-term basis to home buyers and large corporations.
- Savings and loan associations generally serve individual savers and residential mortgage borrowers.
- Life insurance companies receive savings in the form of annual premiums and invest them in stocks, bonds, real estate, and mortgages.
- Pension funds are savings from individuals, corporations and other entities that are invested primarily in the stock market.
- Investment banking houses facilitate the transfer of outstanding securities from one investor to another through markets such as New York Stock Exchange (NYSE) and American Stock Exchange (AMEX).

The goals of efficient intermediaries are:

- Reducing the costs of transferring funds from those with excess funds to those who need funds
- Decreasing risks faced by savers
- Increasing the liquidity of savers

Development of Finance Policies and Procedures

In the 1930s, radical changes occurred when an unprecedented number of business failures caused finance to focus on bankruptcy, reorganization, corporate liquidity, and governmental regulation of the securities market. Through the SEC Act of 1933, the U.S. Securities Exchange Commission (SEC) was formed to ensure that the firms offering securities to the public for purchase provide accurate information about the financial condition of the firm. This Act prohibits the offer or sale through the use of the mails or any means of interstate commerce of any security unless a registration statement is filed. The registration statement is a complete disclosure to the SEC of all material information with respect to the issuance of the specific securities. The SEC Act of 1934 established quarterly, annual and special events reporting requirements for companies who have publicly traded securities.

Prior to the 1950s, a company's financial reporting was primarily focused from the internal point of view to the outside investing community. In the 1950s a new concern developed which focused on the viewpoint of the outside investors to the company's internal financial results. The new focus, assisted by computers, began using complex financial models to analyze cash flows, inventory costs, validity of accounts receivables, return on fixed assets, and other measures of the financial position of the firm.

In the 1960s, companies, attempting to achieve an optimal mix of securities and cost of capital, began to focus on the liability/equity side of the balance sheet as well as on the asset side.

The 1980s witnessed the challenges of dealing with inflation, high interest rates, corporate takeovers, and deregulation of the capital markets. The financial environment created the need for the development of new financial products and structural changes of older financial products. New products included money market funds and the interest rate futures markets. Major structural changes of older products included changing securities from fixed to floating bond rates and variable interest rates.

The late 1980s and early 1990s witnessed a significant recession with interest rates lower than in the previous 30 years, globalization of the economy with domestic firms downsizing, and cutting costs to remain competitive internationally.

All of these evolutionary changes have greatly increased the importance of financial management. In earlier times, the marketing manager would project sales, the engineering and production staffs would determine the assets necessary to meet these demands, and the financial manager's job was simply to raise the money needed to purchase the plant, equipment and inventories. This mode of operation is no longer appropriate. Today, decisions are made in a much more coordinated manner, with the financial manager sharing responsibility for the control process with operating managers.

Multinational Financial Management

As businesses grow, so does their awareness of opportunities in foreign markets. Initially, they may merely attempt to export a product to a particular country or import supplies from a foreign manufacturer. Over time, however, many recognize additional foreign opportunities and eventually evolve into multinational oil companies (MNOCs).

While multinational financial management has been necessary for MNOCs for several years, it is not an exact science. There is still much to be learned. Even the smaller companies are now realizing the need to understand the implications of international finance since international business is not necessarily restricted to the large corporations. If government restrictions do not become excessive, international business should continue to grow. Accordingly, all the financial management decisions related to an MNOCs business, such as financing, working capital management, capital budgeting, and country-risk assessment, will become more critical to their survival and performance.

Also in the 1990s, international oil companies have become more financially stringent in regard to their investment decisions in developing countries. The governments of these countries have reduced their financial support for petroleum development and supply. As a result, international oil companies have become much more selective in undertaking investments and have sought to share project risks by involving other parties, particularly local partners. These preferences have led to more complex financing arrangements addressing the constraints of country limits or weaker or less creditworthy partners. Straightforward corporate financing is no longer fully applicable. Rather, various sources of funds and guarantee instruments are utilized.

INTERRELATIONSHIPS OF OIL PRICES, INTEREST RATES, CAPITAL INVESTMENTS, AND DEBT

In earlier years, the petroleum industry was able to finance a majority of its expenditures through internally generated funds. Debt levels in the industry were minimal until the mid-1980s. In 1983, debt-to-capitalization ratio for a composite of 14 major petroleum companies was 29%. Debt in this equation includes long- and short-term debt, plus production payments, and capital leases. By 1988, debt-to-capitalization ratio for the composite was 41% as detailed in Figure 1–2.

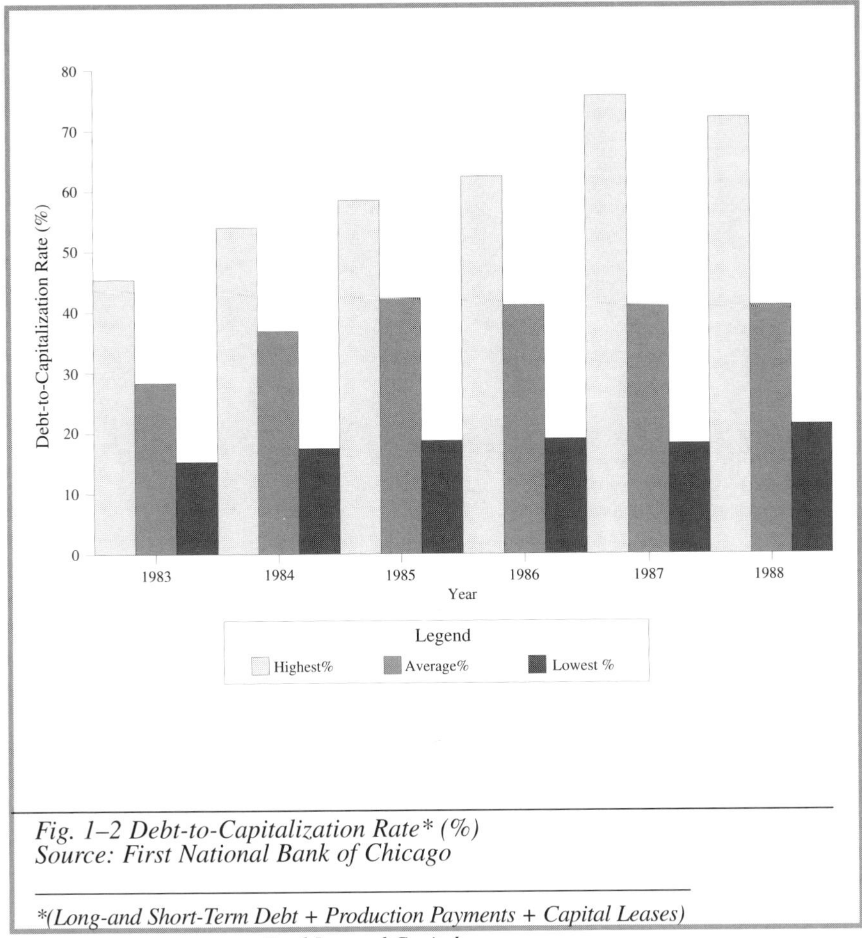

Fig. 1–2 Debt-to-Capitalization Rate (%)*
Source: First National Bank of Chicago

$$\frac{\textit{*(Long-and Short-Term Debt + Production Payments + Capital Leases)}}{\textit{Total Invested Capital}}$$

During the 1980s a primary reason for higher debt levels was the decline in oil prices, which produced a need for external financing to meet operating needs. Beginning in 1984, OPEC attempted to regain market share by driving down the price of oil, forcing some non-OPEC producers out of the market. By 1986, the oil prices had plummeted to $10 per barrel and company earnings closely followed this deterioration. In the 1990s the composite rate decreased with companies reducing debt levels with improvements in the price of oil.

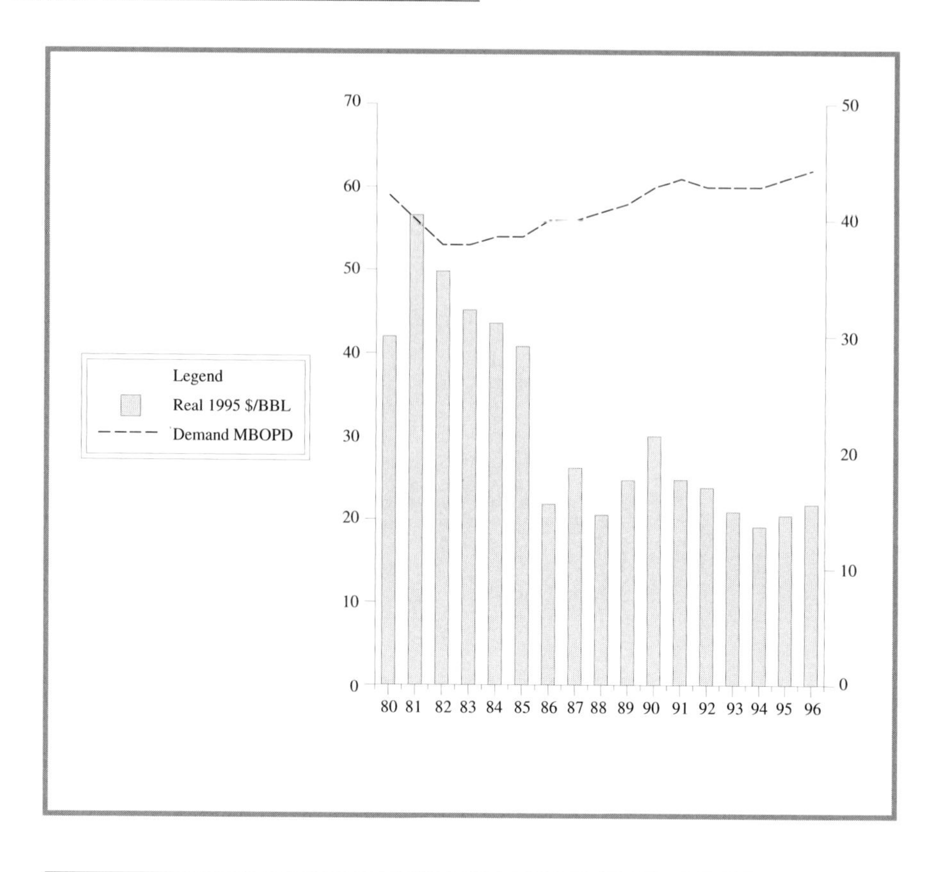

Fig. 1–3 Demand vs. Crude Price in 1995 Dollars
Source: Producer Price Index all Commodities, Bureau of Labor Statistics

Capital expenditures, which historically consumed three-fourths of funds from operations, were cut drastically. Capital expenditure levels are closely related to the price of oil. Projects that had been positively evaluated using a price of crude of $60 per barrel were disbanded or significantly modified. Prior to the 80s, it was believed that the price of finished petroleum products such as gasoline had no sensitivity to the quantity demanded versus a change in price. During the 80s, the question of elasticity was resolved when consumers began to conserve in their usage of gasoline because of the high price. The debate was settled—consumer petroleum product demand is sensitive to price. Automobile owners began new methods of conserving the use of gasoline. Although drastic spending cuts were made, the severe drop in operating funds gave rise to greater external borrowing needs. Therefore, as the price of oil fell the average debt level for the composite increased.

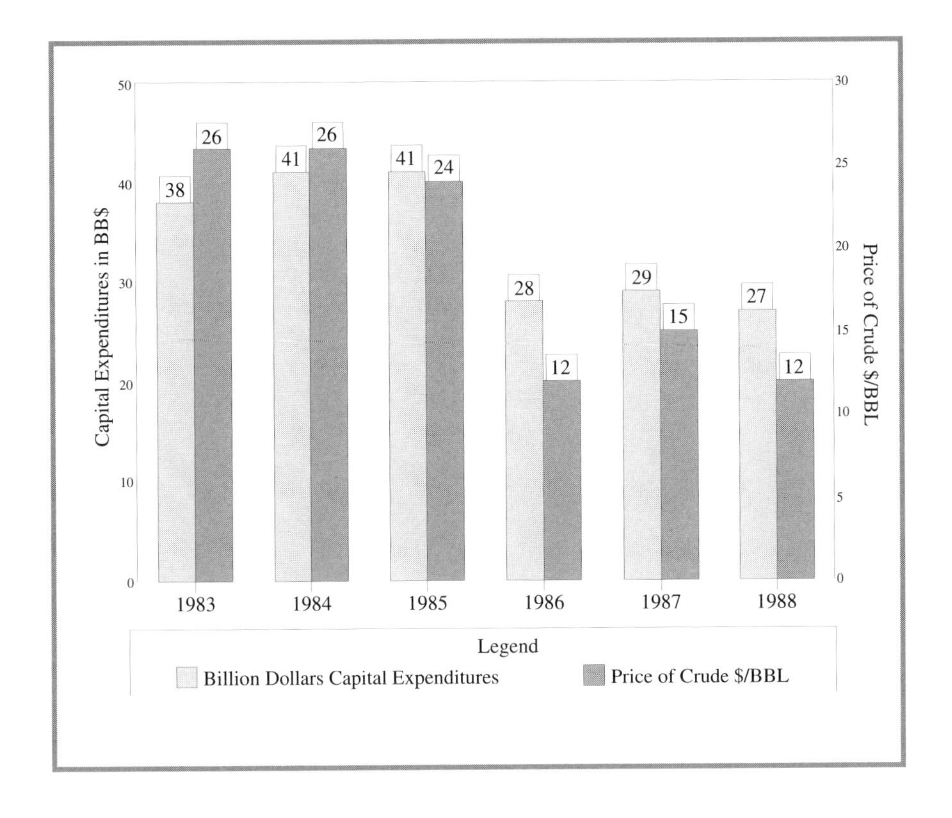

Fig. 1–4 Capital Expenditures in Billion Dollars Compared to Price in $/BBL

In addition to increasing their debt, a number of companies borrowed heavily either to repurchase their own stock or to finance major acquisitions. Stock repurchases were viewed favorably by the investment community in the 1980s as a method to ward off potential corporate raiders. In comparison, such repurchases in the 1960s and 1970s were perceived as a sign of weak management lacking investment opportunities.

In the 1980s, the petroleum industry experienced a number of megamergers, in contrast to smaller diversification efforts in the 1970s. Size, however, was no longer an adequate defense against corporate raiders.

Within two years, debt-to-equity for the composite group grew from 28.3% in 1983 to 42.3% in 1985 as a result of acquisitions and stock repurchases in the industry. Texaco's debt-to-equity increased from 22% in 1983 to 49% in 1984, partially as a result of the $10.3 billion acquisition of Getty Oil. During the same period, Chevron's acquisition of Gulf contributed to its leverage increase from 15.4% to 49%.

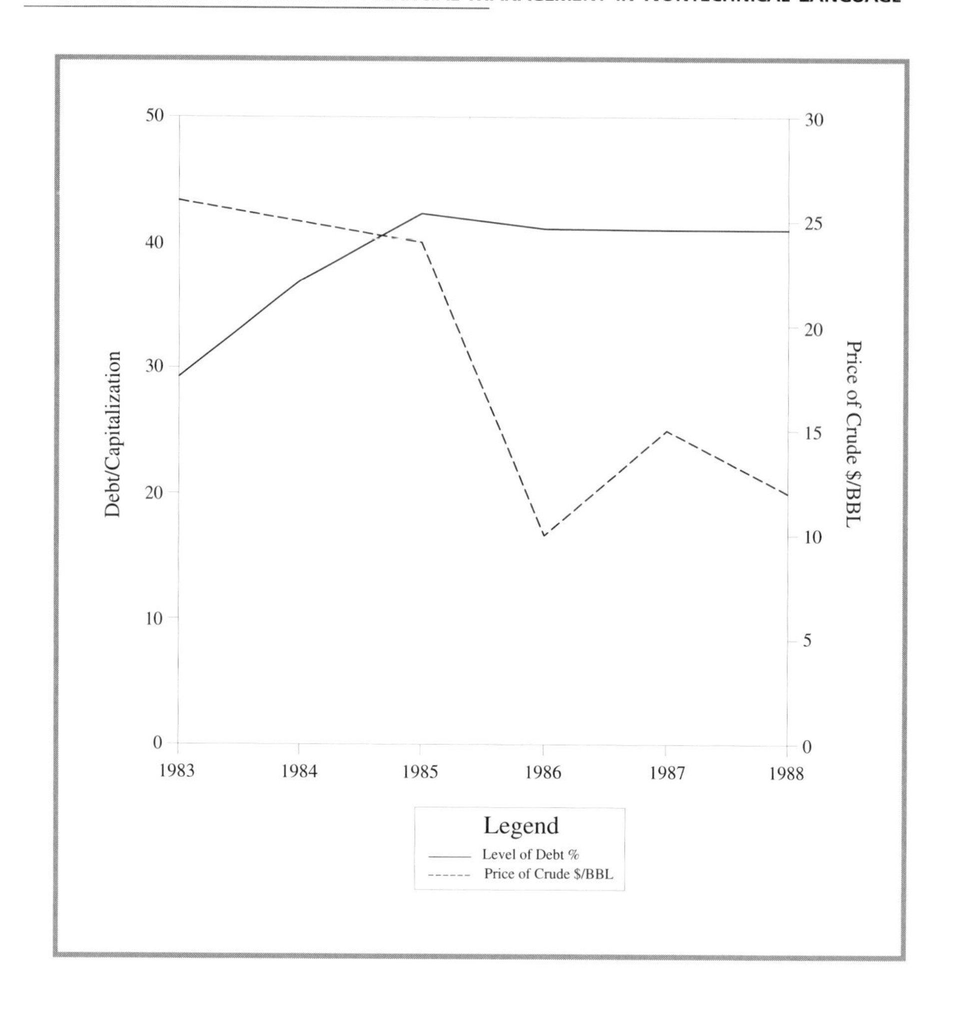

Fig. 1–5 Debt/Capitalization Ratio Compared to Price of Crude

The combination of historically high debt levels and sharply lower oil prices resulted in the deterioration of the petroleum industry's credit condition. In 1986, when the price of oil fell to $10/bbl, interest coverage (earnings before interest and taxes divided by interest charges) for the composite fell to 1.84 times, from 4.3 times in 1985. The bond rating declined from AA+ in 1984 to BBB+ in 1986 for the composite group listed in Figure 1–2.

Although interest rates improved in the mid-1980s through the first part of the 1990s, the interest rate risk became an important issue in the highly leveraged industry. Financial officers are concerned with the amount of floating debt that would be affected if rates rise to the level seen in the early 1980s.

In addition to the predominant risks of oil-price and interest-rate movement, industry members with operations outside the United States are faced with exchange-rate risk. Companies involved only in oil production are not affected as severely by exchange-rate movement, since oil is generally denominated in U.S. dollars. However, local expenditures are denominated in the local currency.

MANAGING RISKS IN THE PETROLEUM INDUSTRY

To meet the needs of the petroleum industry and other industries, the financial markets have developed a variety of products to manage risks associated with oil prices, interest rates,

"Managing the Risk Factors in the Petroleum Industry"

and exchange rates. There are four basic procedures to mitigate risk: forward contracts, futures contracts, options and swaps.

Crude-oil futures and options on the futures have been developed to hedge against adverse oil-price movements. These futures and options are traded on the various worldwide exchanges. This began with the New York Mercantile Exchange (NYMEX) in 1986. However, in the years prior to that, a few firms had already begun offering over-the-counter (OTC) crude options.

Forward contracts are agreements in which a producing company will sell some or all of its future oil entitlement. The oil may be the production from a specified field, or its source may be unspecified. The volume may be specified or unspecified, and the period covered by the contract may be short, long, or unspecified. The buyer may be an end-user with refineries and marketing outlets and their objectives are to secure a reliable supply of crude oil. The buyer may be a trader who is a financial intermediary and their motives will be either speculative or to balance a similar transaction where the intermediary is the seller. Oil and gas forward contracts are not traded on regulated commodity exchanges. The contracts are privately negotiated.

Futures contracts are standardized contracts traded on a commodity exchange. The buyer and seller do not need to know each other's identity because the exchange's rules are structured to provide the necessary financial security. Because the contracts are in a standard form, they are easily traded and highly liquid. The purpose of the futures market is to fix (or speculate in) price, not to secure volume. Some of the exchanges that currently trade energy futures contracts include the New York Mercantile Exchange, The International Petroleum Exchange in London, the Singapore International Monetary Exchange, and the Rotterdam Energy Futures Exchange.

Options include both *sells* (puts) and *calls* (purchase). A call option for petroleum is the right to buy oil (whether futures or physical) at a specified price by a certain date. Conversely, a put option is the right to sell oil (futures or physical) at a specified price by a certain date. The product specified in an option is referred to as the *underlying instrument* or *underlying commodity*. The purchase price referenced in a call option or sale price specified in a put option is called the *strike price* or *exercise price*. An option contract will also include the established time by which the owner must elect to either use (exercise) or not use the option: the exercise date or maturity.

The most commonly used interest-rate option is the cap. Interest-rate caps limit how high the rate can go on a debt instrument. For a cap, the company pays an up-front fee in exchange for a guarantee that another party will pay the interest cost above the agreed upon level. With a cap the company may take advantage of favorable interest-rate movements.

Swaps are flexible instruments and are nothing more than two parties agreeing to exchange some underlying asset for a specified period of time. An oil swap is a commodity swap, which usually involves an oil producer and oil consumer. A financial institution acts as an intermediary, allowing anonymity for the parties involved. In one scenario, when an oil producer expects oil prices to fall, it may pay a floating price per barrel to the intermediary based on an oil index and receive a fixed price in return.

The oil consumer, which may be a refiner, fearing an increase in oil prices will pay the fixed price and receive the floating price through the bank acting as the intermediary. The producer is able to fix the price of the oil it will produce, protecting against a decline in the spot price of oil. The counterpart, the refiner, is protected from a rise in the spot price of oil. Both parties give up their opportunity to benefit from favorable oil-price movement in return for minimizing their respective input or output oil-price risk.

The basic interest-rate swap enables the company to exchange floating-rate payments for another party's fixed-rate payments. The primary reason for interest rate swaps is to change the type of risk and reduce the cost of financing. Typically, two parties want to borrow (or have borrowed) in two different markets. At least one of the borrowers can obtain better pricing than the other in one of the markets. The two markets are typically fixed rate and floating rate. By entering into a swap agreement, both parties can obtain the kind of financing they prefer, while simultaneously taking full advantage of their relative borrowing efficiencies. There is credit risk exposure for a firm entering into interest rate swap arrangements, but it can be minimized through selection of strong financial intermediaries or counterparties if it is a direct arrangement. The risk is also reduced because it does not involve repayment of principal, but only exchange of interest payments.

In currency swaps, each party will have good access to financing in its own currency markets. A typical transaction involves one party that has access to dollar-based financing, but would prefer financing in another currency–Swiss francs, for example. Another party may have excellent access to Swiss franc financing or have large inflows from operations in Swiss francs, but would prefer dollar-based financing

These two parties may both benefit from a swap arrangement. The primary basis for the currency swap is that the two parties can borrow the currencies they need more efficiently (less expensively) through the swap than they can through directly accessing the foreign currency and money markets.

Risk Management and Accidental Losses

The petroleum industry has great concern over environmental losses. As a response to citizen's concerns, governments of most countries have passed considerable and significant legislation requiring certain specifications be followed to protect the environment. Many petroleum companies, in response to their stakeholders, have gone further than the law requires to achieve their goal of being a good citizen.

Some oil companies have formed internal environmental teams composed of environmental engineers, legal, and audit employees. The teams not only define requirements, but also perform environmental audits as part of internal control procedures, and make recommendations to correct any deficiencies. In spite of their preventive measures in efforts and expenditures, catastrophes occasionally occur. It then becomes a question as to how the losses will be covered. This is where the financial manager is involved.

Financial management often has diverse but important responsibilities related to the cost of risk. A primary decision is to identify the potential losses for which the firm should purchase insurance and those for which it would pay for the loss itself or self-insurance. Ultimately, the overall intent is to control the cost of risk from a given set of loss exposures.

The possibility of loss has three broad implications for the firm's financial management:

1. The threat of accidents, even if they do not occur, imposes a very real "cost of risk" on every firm
2. Responsibilities for controlling and limiting accidental losses that may strike the firm
3. Any accidental losses require the firm to have resources available to finance a recovery through either insurance or self-insurance

SUMMARY

Volatile oil prices and currency exchange rates, concurrent with historically high levels of debt, have created new challenges for the petroleum industry. Oil prices in current dollars have ranged from a high of $27/BBL in 1983, to a low of $10/BBL in 1986, with an average of approximately $17/BBL in the 1990s. These fluctuations in price cause major swings in earnings for the industry. In addition, currency exchange-rate movement significantly affects earnings of firms with operations in both domestic and foreign countries.

In this uncertain financial environment, today's management must be able to balance oil-price, interest-rate and, in some cases, exchange-rate risk in order to maintain access to the capital markets. Fortunately, the financial markets have developed numerous products to help manage these risks.

Management in the oil and gas industry is challenged not only with finding funds to pay for exploration but with other significant issues that increase risk, reduce profits and make increasing shareholder/owner's wealth difficult. Becoming knowledgeable in all the techniques available to combat these issues is critical to the success of the company as a whole.

A focus on the international aspects of the oil industry is also necessary as today's oil industry is focusing on worldwide operations. As such, the financial manager must be aware of the international aspects of financing.

Figure 1–6 Managing Financial Risk Factors in the Petroleum Industry

Oil and Gas Financial Reporting and Accounting Systems

2

Financial statements communicate financial information to those outside an enterprise. These statements provide the firm's history quantified in monetary terms. The financial statements most frequently provided are the (1) balance sheet, (2) income statement, (3) statement of cash flows, and (4) statement of owner's or stockholder's equity. In addition, footnote disclosures are an integral part of each financial statement. Financial statements are prepared for creditors and the stockholders—either for judging a company's debt capacity or to assist in determining its stock value.

A company's balance sheet measures the performance of funds deposited by investors in the company (debt as well as equity). It is a detailed presentation of the basic accounting equation: *Assets = Liabilities + Owner's Equity*. The equation is based on the proprietary theory, i.e., the owner's interest in an enterprise (residual equity) is what remains after the economic obligations (liabilities) of the enterprise are deducted from its economic resources (assets). Whether such capital translates into value depends on management's success in earning a high enough discounted cash flow rate of return on that capital. This is the question that, although critical to the economic model of value, no balance sheet can answer. This judgment is best left to the stock market.

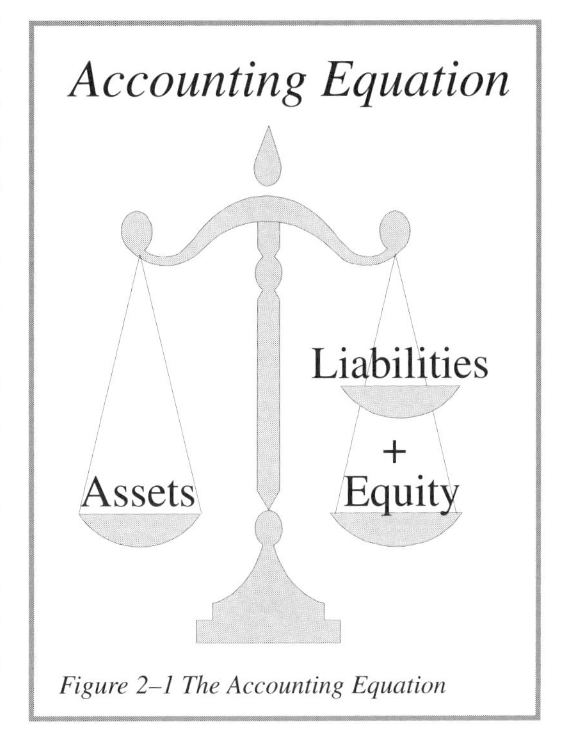

Figure 2–1 The Accounting Equation

Results of operations are reported in the income statement on an accrual basis using an approach oriented to historical transactions. The basic equation is *(Revenue - Expenses) + (Gains - Losses) = Income or Loss*. The current operating performance concept emphasizes the ordinary, normal recurring operations of the entity during the current period. Prior period adjustments are reported in the retained earnings section of the balance sheet.

The *statement of cash flows* provides relevant information about cash receipts and payments of an entity during a given quarter or year. A secondary purpose is to provide information about investing and financing activities. Net income from operations is adjusted by noncash items which are deducted for net income purposes. The most common example of a noncash item is depreciation, depletion and amortization (DD&A).

In addition, management's discussion and analysis addresses liquidity, capital resources, results of operation, and the effects of inflation and changing prices.

The financial statements of a publicly traded company are accompanied by the *report of the independent external auditors*. Their audit is conducted in accordance with generally accepted auditing standards and is intended to provide assurance to creditors, investors, and other users of financial statements.

Objectives of Financial Reporting

In an attempt to establish a foundation for financial accounting and reporting, the accounting profession has identified a set of objectives for financial reporting by business enterprises. Financial reporting should provide the following:

- Material useful to present and potential investors and creditors and others in making rational investment, credit, and similar decisions
- Information to help present and potential investors, creditors, and others in assessing the amounts, timing, and uncertainty of future cash receipts
- Information about the effects of transactions, events, and circumstances upon the resources of the firm

The emphasis on developing cash flows for potential projects might lead one to suppose that the cash basis is preferred over the accrual basis of accounting, i.e., recognition of expenses and revenue before cash being paid or received. This is not the case. Accrual accounting generally provides a better indication of an enterprise's present and continuing ability to generate favorable cash flows.

USERS OF FINANCIAL STATEMENTS AND THEIR NEEDS

Users of financial statements may directly or indirectly have an economic interest in a specific business:

- Users with direct interest are those who make decisions affecting the internal operations of the business and include:
  1. *Investors* or *potential investors* who need information to decide whether to increase, decrease, or obtain an investment in a firm.
  2. *Suppliers* and *creditors* who need information to determine whether to extend credit and under what terms.
  3. *Employees* who want financial information to negotiate wages and fringe benefits based on the increased productivity and value they provide to a profitable firm.
  4. *Management* who need financial statements to assess financial strengths and deficiencies, to evaluate performance results and past decisions, and to plan for future financial goals and steps toward accomplishing them.

- Users with *indirect interests* are those who need to determine whether to create, continue, or terminate a relationship with a firm and include:
  1. *Financial advisers* and *analysts* who need financial statements to help investors evaluate particular investments.
  2. *Stock markets* or *exchanges* that need financial statements to evaluate whether to accept a firm's stock for listing or whether to suspend the stock's trading.
  3. *Regulatory authorities* who need financial statements to evaluate the firm's conformity with regulations in regulated industries.

RECOGNITION AND MEASUREMENT CONCEPTS

According to the *revenue recognition principle*, revenue should be recognized when realized or earned. Revenues are *realized* when goods or services have been exchanged for cash or claims to cash. Revenues are *realizable* when goods or services have been exchanged for assets that are convertible into cash or claims to cash. Revenues are *earned* when the earning process has been completed, and the entity is entitled to the resulting benefits or revenues. The two conditions are usually met when goods are delivered or services are rendered; that is, at the time of sale, which is customarily the time of delivery.

As a reflection of the industry's conservatism, expenses and losses have historically been subject to less stringent recognition criteria than revenues and gains. Expenses and losses are not subject to the realization criterion. Rather, expenses and losses are recognized when economic benefits occur as the company operates and does business or when existing assets no longer provide future benefits. An expense or loss may also be recognized when a liability has been incurred or increased without receiving a corresponding benefit. An example is a contingent loss or inflation.

Limitations of Financial Statement Information

The measurements made in financial statements do not necessarily represent the true worth of a firm or its segments:

- The measurements are made in terms of money; therefore, qualitative aspects of a firm are not expressed.
- Information supplied by financial reporting involves estimation, classification, summarization, judgment and allocation.
- Financial statements primarily reflect transactions that have already occurred; consequently, they are usually based on historical cost.
- Only transactions involving an entity being reported upon are reflected in that entity's financial reports. However, transactions of other entities, like competitors, may be very important.

Historical Cost Limitations

Most transactions reported by financial statements are recorded at their value on the date of the transaction. Assets previously acquired are recorded on the balance sheet at their historical cost. When the values of these assets change significantly after the acquisition date, the balance sheet presentation of the assets becomes significantly less relevant in determining the company's worth. Over time, discrepancies develop between current and historical values of a transaction or an asset. For example, the replacement cost and the book value of assets will diverge. Moreover, the value of the unit of measure (the dollar for U.S. firms) will also change. Accordingly, comparisons between prior years and between competing firms become less meaningful unless adjusted for price level changes.

Realization Concept

The *realization principle* dictates that revenue should be recognized only at the time a transaction is completed with a third party, or when the value is reasonably certain. One common concern is applying the realization principle to the petroleum industry. Some feel this principle should not apply because the major asset of an oil and gas company is its reserves, and the value of a company's reserves is not directly reflected on the balance sheet.

Neither the balance sheet nor the income statement allow appropriate recognition for important oil and gas discoveries in the accounting period in which a discovery is made. When a company makes a major discovery, there is no mechanism for reporting the results from an accounting point of view. The impact on the income statement comes when the discovery begins to produce, yet economic value is realized the moment the discovery was made. But just how much economic value? This is the first natural question posed by the accountant. It is a fair question, too, because at the point of discovery, the uncertainty as to the quantity and value of reserves is greatest. Fortunately, the analysis of a company does not end with the financial statements.

The Matching Concept

The *matching principle* provides that revenues should be matched with the corresponding costs of producing such revenues. A serious accounting issue is the matching principle because it is difficult to match the costs of finding oil and gas with the revenues from production. Under General Accepted Accounting Principles (GAAP), assets reported on the balance sheet consist of capitalized historical costs. Earnings are recognized as reserves are produced, rather than when they are discovered or revised.

Two separate systems of accounting in the industry are based primarily on this issue as it pertains to the treatment of exploration costs. The two systems are called *Full Cost Accounting* (FC) and *Successful Efforts Accounting* (SE). While revenues are typically recognized when oil or gas is sold, the fundamental difference between the FC and SE accounting systems lies in how the corresponding costs of finding reserves match the revenues.

The most valuable assets of an oil and gas company are the reserves not recorded in the accounting system. To provide oil and gas reserve data and other information not recorded in the financial accounts, the FASB and SEC have developed procedures for reporting this data in a supplemental section of the financial report. FASB Statement No. 19 provides reporting requirements for the supplemental information frequently called Reserve Recognition Accounting (RRA). Chapter 10 discusses RRA in detail.

The Cost Concept

In accounting, an asset is recorded at its original cost. This cost is the basis for all subsequent accounting for the asset. The primary rationale behind the *cost principle* is that the value of an item may change with the passage of time, and determination of value is subjective. There is no subjectivity associated with the actual cost of an item.

Because of the cost principle, the accounting entry on the balance sheet for oil and gas assets usually has little to do with the actual value of the assets. For instance, if a company were to obtain a lease and then discover a million barrels of oil, the accounting entry would not change because of the discovery. It would never reflect anything other than the associated costs less depreciation, depletion and amortization (DDA). A conservative argument is "how valid is the quantification of the discovery?"

In this example, there is substantial appreciation in value that is ignored by GAAP. However, accountants do not ignore economic value completely. The cost principle provides that assets should be reflected on the balance sheet at cost, unless there has been a decline in their utility or economic value. Accountants do not mind an asset on the books at less than its true market value, but they are careful to keep accounting entries from exceeding original costs.

COST CLASSIFICATION

Costs associated with oil and gas exploration and production fall into four fundamental categories:

- *Lease Acquisition Costs*. Costs associated with obtaining a lease or concession and rights to explore for and produce oil and gas.
- *Exploration Costs*. Costs incurred in the exploration for oil and gas such as geological and geophysical costs (G&G), exploratory drilling, etc.
- *Development Costs*. Costs associated with development of oil and gas reserves. Drilling costs, storage and treatment facilities, etc.
- *Operating Costs*. Costs required for lifting oil and gas to the surface, processing, transporting, etc.

Treatment of these costs is fairly straightforward. The one exception is the way that exploration costs are treated. This provides the basis for the two different accounting practices, SE and FC, that are used in the industry.

Cost Centers

One of the main differences between SE and FC systems results from the size and use of *cost centers*. It is this difference that makes the financial impact. With SE, costs for a cost center can be suspended until it is determined if commercial quantities of oil or gas are present.

With a well or lease as the cost center, costs are expensed if the well is dry and capitalized if it is a discovery. This can be a very subjective decision. Sometimes the decision to drill a well may be held up because of perceived impact on the financial statements during a specific period, should the well turn out to be unsuccessful.

Under FC, a cost center can be defined as the entire nation or world. Consequently, all costs are capitalized regardless of whether a discovery has been made. The costs would not appear in the income statement but rather on the balance sheet. The amount of net income is better but the return on capital employed suffers.

Ceiling Test Limitation

FC accounting requires a write-down on the book value of oil and gas assets if it exceeds the SEC value of reserves. This is the ceiling test required by the SEC for the cost of oil and gas properties on the balance sheet. The recorded capitalized costs for producing oil and gas properties are limited to the net present value of the reserves discounted at 10%. This is the SEC value of reserves or standard measure.

If the SEC value of reserves falls below the capitalized costs on the balance sheet, a ceiling write-down occurs. For example, if a company had a book value for proved oil and gas properties of $100 million, and the SEC value of these reserves was $130 million, there would be no write-down. Instead, the company would have a cost ceiling cushion of $30 million.

In 1986, when oil prices dropped dramatically, cushions disappeared and many FC companies experienced substantial write-downs. The most important problem that this caused was many companies suddenly found themselves in violation of covenants in their loan agreements.

Impairments and write-downs occur under SE accounting, too. It is usually not considered as great an issue because such a large part of exploration costs are expensed and not capitalized. But, consistent with the conservative principle, the carrying value of SE oil and gas properties are subject to write-downs if the economic value of a property is less than the recorded value. Periodic assessments are made to ensure that the value of leases have not been impaired due to negative results of drilling or approaching expiration dates.

Comparison

With each system, lease bonus payments, related legal costs, and development drilling costs are capitalized. Capitalized costs within a cost center are usually amortized on the unit-of-production method (explained later in this chapter). Basic elements of the two systems are compared in Table 2–1.

	Successful Efforts	Full Cost
G&G Costs	Exp	Cap
Exploratory Dry Hole	Exp	Cap
Lease Acquisition Cost	Cap	Cap
Successful Exploratory Well	Cap	Cap
Development Dry Hole	Cap	Cap
Successful Development Well	Cap	Cap
Operating Costs	Exp	Exp
Which Companies Typically Use Each Method	Major Oil Companies	Smaller Independent Companies
Size of Cost Center Used	Small	Large
	Single Well, Lease or Field	Company, Country, or Hemisphere
Comment	Favored by FASB Approved by SEC	Approved by SEC

G&G = Geological and Geophysical
Exp = Expensed (Written off in accounting period)
Cap = Capitalized (Written off over a number of accounting periods)

Table 2–1 Comparison of Successful Efforts and Full Cost Accounting Methods

Book Value, SEC Value, and Fair Market Value of Reserves

The SEC value of oil and gas reserves that appears in the financial reports bears no direct relationship to reserve values under either FC or SE accounting methods. Because the SE company will expense G&G costs and exploratory dry holes, the book value of reserves will be lower than under FC accounting. The SEC value of reserves is usually less than the market value for reserves. Analysts must look beyond the reported figures on the balance sheet and the SEC values of reserves to determine economic value.

Comparison of Accounting Impact for a Startup Company

Table 2–2 illustrates an example of a company with $10 million startup capital. In its first year, the company drills 15 wells and has two discoveries. The following table outlines the general features of how the two accounting methods would report the financial results of the first year of operations.

Exploration Wells Drilled	Drilling Costs ($000)	Results
6	3,000	Plugged and Abandoned
1	1,500	700,000 Barrel Discovery
7	2,500	Plugged and Abandoned
1	1,000	900,000 Barrel Discovery
15	$8,000	1,600,000 Barrels

Generalized Financial Reporting

	SE Accounting ($000)	FC Accounting ($000)
Revenue	0	0
Expense	(5,500)	0
Income*	(5,500)	0
Assets		
Cash	2,000	2,000
Property	2,500	8,000
Equity	4,500	10,000

*No income taxes or G&A costs considered.

Table 2–2 Company Startup Results Under Full Cost and Successful Efforts Accounting

Under the SE method, all exploratory dry holes are expensed. For the startup company, reporting a $5.5 million loss in the first year can be devastating. This is one of the reasons why so many small startup companies prefer FC accounting. The company managed to find 1.6 million barrels of oil. The company might believe the reserves would be unfairly represented by a book value of only $2.5 million under SE.

Events affecting net income will also be represented differently under the two accounting systems. This is particularly true of a startup company. With moderate reinvestment of cash flow, the earnings of an established company would be nearly the same under either accounting system.

Figure 2–2 depicts how the income statement under the two accounting systems would reflect changes in drilling activity. It is assumed that drilling efforts result in the same degree of success as in the past. With increased exploratory drilling, net income drops under SE accounting compared to FC accounting methods. This is because SE will expense exploration dry-hole costs. Decreased exploratory drilling under both methods will increase net income initially, but the change will be greater under SE.

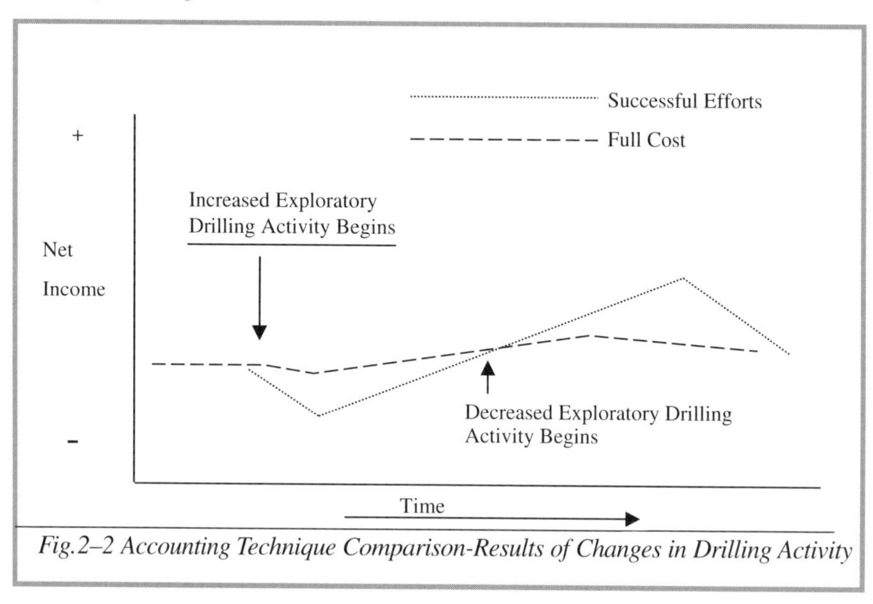

Fig.2–2 Accounting Technique Comparison-Results of Changes in Drilling Activity

The comparison in Figure 2–3 illustrates the effect on earnings of an increased rate of discovery, that is, a greater percentage of successful wells rather than dry holes. Under FC accounting, the impact on net income is shown in later years when the additional discovered oil begins to come on stream. The difference with this scenario is that fewer dry holes are written off for the SE company. The FC company would capitalize exploration wells whether successful or dry.

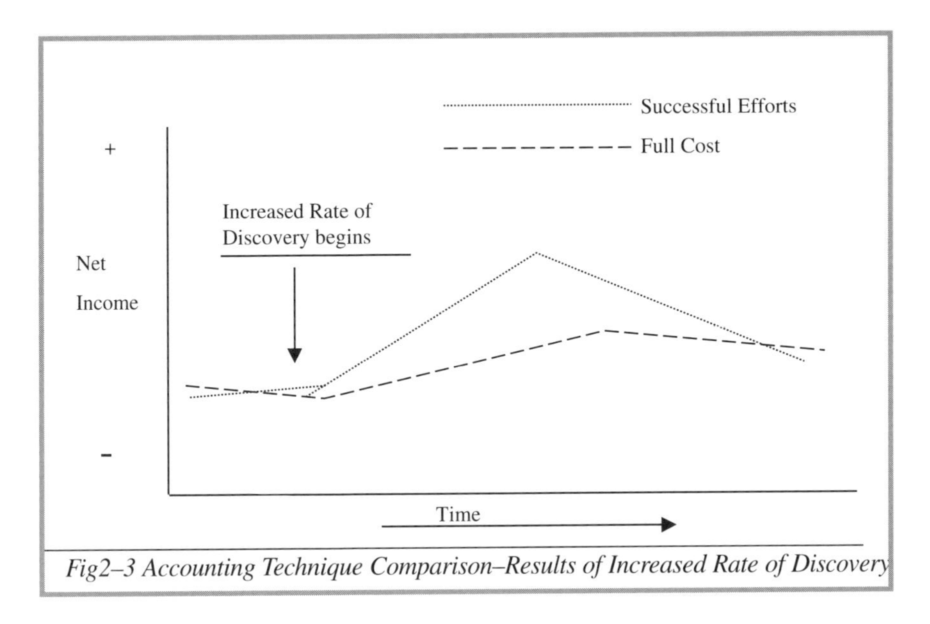

Fig2–3 Accounting Technique Comparison–Results of Increased Rate of Discovery

For the SE company, if greater success is the result of more reserves per well as opposed to a larger ratio of successful wells, then the result may be more like that of an FC company. The difficulty with both SE and FC accounting methods is that increased success in exploration is not shown until subsequent years. It is helpful that the company's net reserves are reported each year so analysts can know at least whether reserves are being increased or depleted.

The SEC value of reserves helps quantify the company's success in not only replacing reserves but also replacing value. It would not be a good situation for a company to produce and sell high-quality reserves and replace them with low-quality, low-value reserves.

Depreciation, Depletion, and Amortization

Depreciation is a means of accounting for the recovery of the costs of a fixed asset by allocating the costs over the estimated useful life of the asset. When this concept is applied to mineral resources such as oil and gas reserves, it is called *depletion*. The concept is called *amortization* when the allocation of costs is applied to intangible assets. The terms *depreciation, depletion,* and *amortization* (DD&A) are sometimes used interchangeably, or more often collectively, as DD&A. The importance of DD&A is that these expenses are deducted from income for federal and state income tax purposes. The depreciable life of an asset is usually determined by legislation to emulate the useful life of that asset. Table 2-3 provides examples of typical lives of oil industry assets.

Asset Category	Asset Lives
Intangible Drilling Costs that must be Amortized (30% of IDCs)	5 Years
Vehicles and Drilling Equipment	5 Years
Production Equipment and most Field Equipment Office Equipment Processing Plants	7 Years
Refining Equipment	10 Years
Transmission Pipelines	15 Years
Buildings	30 Years

Table 2–3 Typical Asset Lives for Depreciation and Amortization

Cost Depletion and Percentage Depletion

Depletion for tax purposes is based on the concept of the removal and sale of a wasting or depleting asset—in this case, oil and gas. Depletion is analogous to ordinary depreciation. The Internal Revenue Code authorizes a deduction from income for depletion of oil and gas properties. It is a relatively simple concept, but it is complicated by many limitations and exemptions.

Two methods of calculating depletion must be considered when estimating taxable income. The two methods are *cost depletion* and *percentage depletion*. The taxpayer is entitled to the higher of either the cost or percentage depletion. Cost depletion of producing oil and gas properties is allowed under the unit-of-production method on a property-by-property basis. This method requires a reasonably accurate estimate of the remaining recoverable reserves. The basic formula is outlined in Figure 2–4.

The tax advantages of percentage depletion were eliminated for all integrated oil companies in 1975. Cost depletion is available to both independent producers and integrated oil companies. Usually the percentage depletion allowance is higher than cost depletion.

Percentage depletion allows a producer to deduct from gross income a stated percentage of gross income as an expense. Originally, in 1926, the Internal Revenue Code for oil and gas wells allowed a deduction of 27.5% of gross income from production. This method of depletion was very controversial and is no longer granted to the major oil companies, but independent com-

Annual Depreciation = *(C - AD - S) P÷R* where:

C = capital costs of equipment
AD = accumulated depreciation
S = salvage value
P = barrels of oil produced during the year*
R = recoverable reserves remaining at the beginning of the tax year

*If both oil and gas production are associated with the capital costs being depreciated, then the gas can be converted to oil on a thermal basis (6 MCF of gas equals one barrel of oil).

Figure 2–4 Formula for Unit-of-Production Method

panies are still allowed a percentage depletion allowance.

The percentage depletion allowance since 1984 has been set at 15% of gross income. The depletion allowance is not to exceed 65% of net taxable income (computed without including depletion allowance). Prior to 1975, the limit was 50%.

The combined DD&A for many companies can be quite significant. The per-unit values that are deducted for income tax purposes can range from $3 to over $10 per barrel. The majority of oil and gas property valuations in the United States are done on a before-tax basis.

CASH FLOWS

Accrual vs. Cash

Another aspect of the realization concept is the *accrual method* accounting for revenue and expenses. Under this method, revenue is recorded as it is earned, or is said to have "accrued", and does not necessarily correspond to the actual receipt of cash. This concept is important for the understanding of the statement of cash flows and the concept of cash flow as illustrated later in Table 2–6.

The income statement would reflect $20,000 because the accrual method of accounting realizes the income at the point of sale, not at the point of actual cash exchange. The balance sheet will show the $17,000 increase in cash as well as an increase in account receivables of $3,000 not yet received.

Assume that a company sold 1,000 barrels of oil for $20 per barrel, but had only received $17,000 by the end of the accounting period. From an accrual accounting point of view, revenues are recorded as $20,000.

Revenues	20,000	
Beginning Receivables	1,000	
Cash Flow Potential	$21,000	
Ending Receivables	-4,000	
Realized Cash Flow	$17,000	= Sales less increase in receivables

However, the actual cash received is $17,000. This is why the statement of cash flows treats increases in the working capital asset accounts as a reduction in cash flows. As a business grows, the required amount of working capital also increases, and therefore, most detailed cash flow analyses include a negative adjustment for increases in working capital.

"Oscar! Would you please tell me why we don't have any cash and your reports indicate income????"
"But Boss! The accounts receivable increased and the accounts payable decreased!!!"

EARNINGS VS. CASH FLOW

Exploration, development, and production operations budgets, funded by cash flow, are often substantially greater than company earnings. Also, the investment community does not overlook the importance and superiority of cash flow compared to earnings. Cash is king. Earnings and earnings multiples are quoted frequently because of the accessibility of these figures. But most analysts focus on cash flow. The accounting methodology used in the oil industry can yield very different results on reported earnings depending on the method used.

Companies that use SE accounting methodology will expense unsuccessful exploration costs. On the other hand, companies that use FC accounting will capitalize these costs and write them off over a number of accounting periods.

By adding exploration expenses and DD&A to net income, the differences between SE and FC accounting begin to disappear. Cash flow analysis then begins to place companies on an equal footing for comparative purposes. A comparison is made of two otherwise identical companies under SE and FC accounting. The simplified income statements of the two companies in Table 2–4 illustrate the importance of cash flow.

	Company A Successful Efforts, $	Company B Full Cost, $
Revenues	10,000	10,000
Expenses	7,000	7,000
Exploration expense	1,000	-0-
DD&A	1,500	2,000
Net Income	500	1,000
Income Tax (34%)	170	340
Net Income after tax	330	660
Plus DD&A	1,500	2,000
Plus Exploration Expense	1,000	1,000
Conventional Cash Flow	2,830	2,660
Income Tax	170	340
Before-Tax Cash Flow	$3,000	$3,000

Table 2–4 Comparison of Earnings and Cash Flow

The earnings for Company B are twice that of Company A, but Company A has more cash flow. The differences in accounting for depreciation will impact the taxes paid. Using before-tax net income is one way to place the two companies on a more equal footing for comparison. An important thing to consider here is that not all SE companies treat dry hole costs in the same manner on the statement of cash flows. About half of the SE companies will add dry hole expenses as an adjustment to earnings in calculating cash flow from operations (CFFO). The other half do not.

Calculating Cash Flow

Sometimes cash flow calculations may have adjustments that are specific to an industry or to a particular situation. A detailed cash flow analysis may include assumptions about changes in corporate strategy. Certain expenditures such as research and development expenses might be considered unnecessary, especially in a short-term financial crunch.

Most analysts start with net income and make the necessary adjustments to arrive at cash flow. The main difference between cash flow and net income is in the adjustments that are made for noncash and nonrecurring items. Examples of the most common elements of cash flow are:

- Extraordinary items that include loss on sale of assets, write-downs on book value of assets, and gain on sale of assets
- Exploration expenses
- Depreciation, depletion, and amortization (DD&A)
- Interest expense

Cash Flow Statements

Beginning in 1988, a statement of cash flows (SCF) was required by the Financial Accounting Standards Board (FASB), instead of its predecessor the statement of changes in financial position (SCFP). SCFs became effective for financial statements for fiscal years ending after July 1988. The SCF segregates information about cash provided or used by a company into three categories:

1. Operating activities (cash flow from operations)
2. Investing activities
3. Financing activities

The information included in the SCF for a company helps determine its

- Ability to generate future cash flows
- Capacity to meet financial obligations
- Success in investing strategies
- Effectiveness in financing strategy

The Arithmetic of the SCF is outlined in Table 2–5.

Cash Flows from operations	= Income net of interest and taxes
	+ DD&A
	+ Deferred taxes
	+ Losses on asset sales
	- Gains on asset sales
	- Increases in working capital or
	+ Decreases in working capital
Cash flows from investing activities	- Payments to purchase property, plant, and equipment (PP&E) and other productive assets
	+ Receipts from the sales of PP&E and other productive assets
Cash flows from financing activities	+ Proceeds from issuance of equity instruments, securities and notes payable
	+ Proceeds from issuing bonds and other short- or long-term borrowing
	- Payments of dividends, including treasury stock purchases
	= Net increase (decrease) in cash and cash equivalents

Table 2–5 Statement of Cash Flows—Arithmetic

Discretionary Cash Flow

Discretionary cash flow frequently represents internally generated cash flow. Out of these funds, dividends can be paid and exploration and development ventures undertaken. Similar definitions do not consider preferred or common stock dividend payments a discretionary item, and they are deducted. The rationale behind this is that management really does not have much latitude to consider dividend obligations a truly discretionary matter.

The common definition of cash flow comes in various forms and disguises. Usually the terms cash flow and discretionary cash flow are used interchangeably. Table 2–6 contains a collection of definitions that address a form of cash flow, but the terms and definitions appear dissimilar. With the exception of EBITXD, the terms and definitions in the box are consistent with what is usually called *discretionary cash flow*. However, while the simple definition of discretionary cash flow is useful and quick, there are more elaborate definitions.

- **Discretionary income** (Enron Oil & Gas, 1989 Annual Report) is net income (loss) adjusted to eliminate the effects of DD&A, lease impairments, deferred taxes, property sales, other miscellaneous noncash amounts and exploration and dry hole expenses.
- **Discretionary cash flow** (Maxus Energy Corp., 1990 Annual Report) is net income (loss) plus noncash items and exploration expense, reduced by preferred dividends.
- **Discretionary cash flow** (Oryx Energy Co., 1990 Annual Report) is the cash from operating activities, before changes in working capital, plus the cash portion of exploration expenses.
- **Operating cash flow** (Snyder Oil Co., 1990 Annual Report) is net income before DD&A and deferred taxes.
- **Cash from operations** (Apache Corp., 1990 Annual Report) is cash flow from operating activities before changes in components of working capital.
- **Working capital provided by operations** (Berry Petroleum Co., 1990 Annual Report) is effectively cash flow from operations (CFFO) before changes in working capital.
- **Cash flow before balance sheet changes** (Devon Energy Corp., 1991 Annual Report) is CFFO before changes in components of working capital.
- **Net operating cash flow** (Edisto Resources Corp., 1991 Annual Report) is net loss as adjusted for depreciation, depletion, amortization, abandonment, and exploration costs, plus losses on asset sales.
- **Operating cash flow** (Sanford C. Bernstein & Co. Inc.) is net operating income plus DD&A plus deferred taxes plus exploration expense.
- **Cash flow** (Wertheim Schroder & Co., Inc.) is adjusted net income plus DD&A plus after-tax exploration expense (which was determined by applying actual or statutory tax rates to the current year change).
- **EBITXD** (First Boston) are earnings before interest, taxes, exploration expense, and DD&A.

Table 2-6 The Spectrum of Cash Flow Definitions

Free Cash Flow

Cash flow, as most definitions structure it, does not represent the true profitability of a firm much more than earnings. The treatment of depreciation is an important matter, and simply adding DD&A to net income should be considered in the proper light. The thought of a noncash item can be an obstacle because it is slightly abstract, but once the concept is grasped the door is open to an understanding of cash flow theory.

The usual treatment of cash flow, adding all noncash items to net income, incorrectly ignores the need for replenishment of assets. There is sound reasoning behind DD&A. The best definitions of cash flow are those that acknowledge the need for capital to maintain a company as a going concern. Where a company decides to self-liquidate, a pure cash flow analysis that purposely ignores the need for capital infusion is appropriate.

Free cash flow represents cash flow available after the necessary capital expenditures have been made to sustain a company's productive capacity. This treatment is known as maintenance capital and is the fundamental difference between free cash flow and discretionary cash flow. For an oil company, maintenance capital would include funds necessary to drill wells and maintain facilities such as refineries and pipelines.

Most analysts ignore maintenance capital when doing a quick analysis, but they still know the importance of this item. Maintenance capital is virtually the opposite of DD&A. As a company grows, required increases in working capital are also considered a part of maintenance capital.

Cash Flow from Operations

Cash flow from operations (CFFO) is the net amount of cash taken in or lost from operating activities during a specific accounting period. The CFFO appears in the cash flows from operation section of the SCF.

The CFFO figure is sometimes treated as cash flow although it is not the same as the cash flow figures normally quoted. The main difference between CFFO and discretionary cash flow is the treatment of changes in the components of working capital.

Most analysts do not explicitly address the changes in working capital for basic quick-look cash flow analysis. This is one reason why some companies will summarize net cash flows provided by operating activities before changes in components of working capital. This issue involving components of working capital is due to differences between accrual and cash accounting.

SUMMARY

Analysts understand that accounting methods can have a significant impact on reported earnings. Earnings for FC companies are usually considered to be inflated in comparison to SE companies. FC companies pay a price for the opportunity to report relatively higher earnings. They must also pay relatively more in taxes. Yet, ignoring this aspect, the intrinsic value of an oil company is the same regardless of the accounting technique used.

Analysts who look at breakup value of a company ignore the book values that different accounting systems may yield. Analysis of asset values neutralizes the differences between FC and SE accounting methods. The same is true of cash flow analysis. When an analyst looks beyond reported earnings and starts analyzing cash flow, the differences between FC and SE begin to disappear.

It is important to understand the difference between cash flow and net income when the financial manager is planning capital projects or analysts are evaluating the firm.

Integrated Management System

3

Business planning, when done properly with all the functions or departments of the entity participating, is a complicated but fascinating procedure. The idea of business planning is not new. Many companies have used formal or organized planning for years. What is different, however, is the emergence of a new approach to planning and a new sense of urgency about the need to plan. Planning in the 1990s moved out of the corporate planning department and infiltrated all staff and operating departments. Employees at all levels began planning their processes to meet the needs of their customers and to achieve company goals.

The agents of change that, until the 1950s, lumbered along at an imperceptible pace have picked up speed dramatically. Change is constant and the best answer for change is change itself. Until the 1950s, global population and economies were slow and sluggish. However, since then agents of change have taken over. The population has grown faster in the last four decades than in the previous four million years. The global economy has grown sevenfold to nearly $30 trillion.

One aspect of change is less emphasis on company size. Until recently, large was better in the corporate world. Larger companies would have in-house services, robust balance sheets, a huge asset base and access to privileged information. Today, cheap information is available to everyone. Most companies work with a mentality of *just-in-time, customer satisfaction,* and *reengineering.*

Importance of Planning in the Oil and Gas Industry

Since its beginning, the petroleum industry has experienced cycles of economic prosperity and depression. A few of these cycles have lasted for one-half decade. Companies spend a good deal of time planning ahead so they can anticipate the future cycles. If they cannot anticipate the future, at least they can consider alternate scenarios of what might occur in the future and the events that lead to the *end state* (See Scenario Planning).

The lead time between exploration money spent and revenues received from a customer at burner tip or service station is a significantly long time span of up to one or more years. Besides lengthening lead times, the size of investments increase as companies shift exploratory programs to the deeper parts of known producing areas. Exploratory programs have extended to the more hostile offshore areas and to the far reaches of our world, often in uncertain political environments. Everyone in the oil and gas industry recognizes the risks inherent in the search for hydrocarbons. Today, managers are becoming concerned regarding major risks that exist in all of the segments of the oil industry including downstream refinery operations and marketing of either liquid or gas products.

Integrated oil companies have spun off investments to form new joint ventures, alliances and partnerships. Amoco and Shell Oil combined oil and natural gas production operations in West Texas and Southeast New Mexico via a limited partnership, Altura Energy Ltd. The partnership is 64% owned by Amoco and 36% owned by Shell. The new organization expects to save the companies hundreds of millions of dollars during the next 15 years.

Shell and Texaco, seeking to reduce costs, combined refining and marketing operations that accounted for 14% of gasoline sales in the western and Midwestern United States. Shell owns 56% and Texaco 44% of the new company. The merger included all their nationwide transportation and lubricant businesses under the same company.

Unocal sold its interest in UNOVEN, a refining and marketing joint venture with Petroleos de Venezuela SA (PDVSA), to CITGO. PDVSA is the parent company of CITGO. Strategies to reduce competition and improve financial performance are the result of mature oil companies seeking to improve shareholder value and return on capital employed. There is a tendency to be more aware of the need for better management processes when the environment changes. Many companies have integrated their planning, budgeting and reporting systems into a new idea called Integrated Management System.

Processes of the Integrated Management System

The Integrated Management System (IMS) is an integrated group of four management processes that support execution of business strategies. IMS modifies management processes and integrates them with new processes while maintaining a common focus on execution of strategy. The four management processes are *strategic planning, long range financial forecasting, annual performance planning* and *periodic performance reporting.*

Strategic plans define long-range operational goals and broad boundaries of operation for the business. The milestones established within the strategic plan provide direction toward accomplishing these goals. A meaningful strategic plan should build a picture of a company's current and future environment. Then it should address how the business will move from the current scenario to the new scenario. Where will new exploration occur? Which resources are required? Where will new refineries be built? Where will natural gas pipelines be built? Which new and types of markets might be entered for selling petroleum products to the consumer? Which changes in costs or technologies are likely? What competition will be addressed? A financial model flows from the projected future.

Long-range financial forecasts provide an internal check for business plans and a benchmark for company results. A solid understanding of the economy and the business environment helps in setting revenue targets, forecasting cost changes, developing sales forecasts, and selecting markets for new products or company expansion. Several types of data are used in developing a strategic plan. They include economic, demographic and business environment data. The information obtained from a variety of sources include data vendors, consulting groups, university research organizations, and government agencies. Cost, reliability, and reputation within the industry are among the factors to consider in selecting information sources.

Annual performance plans, including operational plans, forecasts, and budgets, are developed based on and in support of strategic plans. An annual performance plan defines business performance in terms of milestones and targets (strategic, operation, financial) met during the year. The milestones agreed upon and defined within the annual performance plan establish *accountability* for achieving *business results.*

Balanced performance reports monitor the results achieved compared with the milestones and show progress toward the long-term goals defined in the strategic plan. They also provide feedback as a basis for taking corrective action and rewarding results.

STRATEGIC PLANNING

Strategic plans are the general long-range plans for the firm's future, often developed on a three-to-five-year basis and revised annually. The basic purpose of the strategic plan is to chart a course for the firm over the long term—more specifically, what it needs to do in the next few years to achieve its long-term objectives. This may involve determining the progression of debt and equity over the upcoming periods and their impact on the firm's financial structure. The developed plan becomes a "road map" for the firm to follow. For the financial planner, strategic plans are important because they establish the basic assumptions necessary for developing more detailed financial plans. Strategic plans usually include the following components:

- *Corporate purpose.* This is a general statement that defines the general purpose of the firm. Usually it includes a "mission statement" that shows the firm's future direction. While defining corporate purpose, the company should consider the stakeholders or those who have expectations from the organization. Things change over time, and the corporate purpose valid today may not be valid in the future.

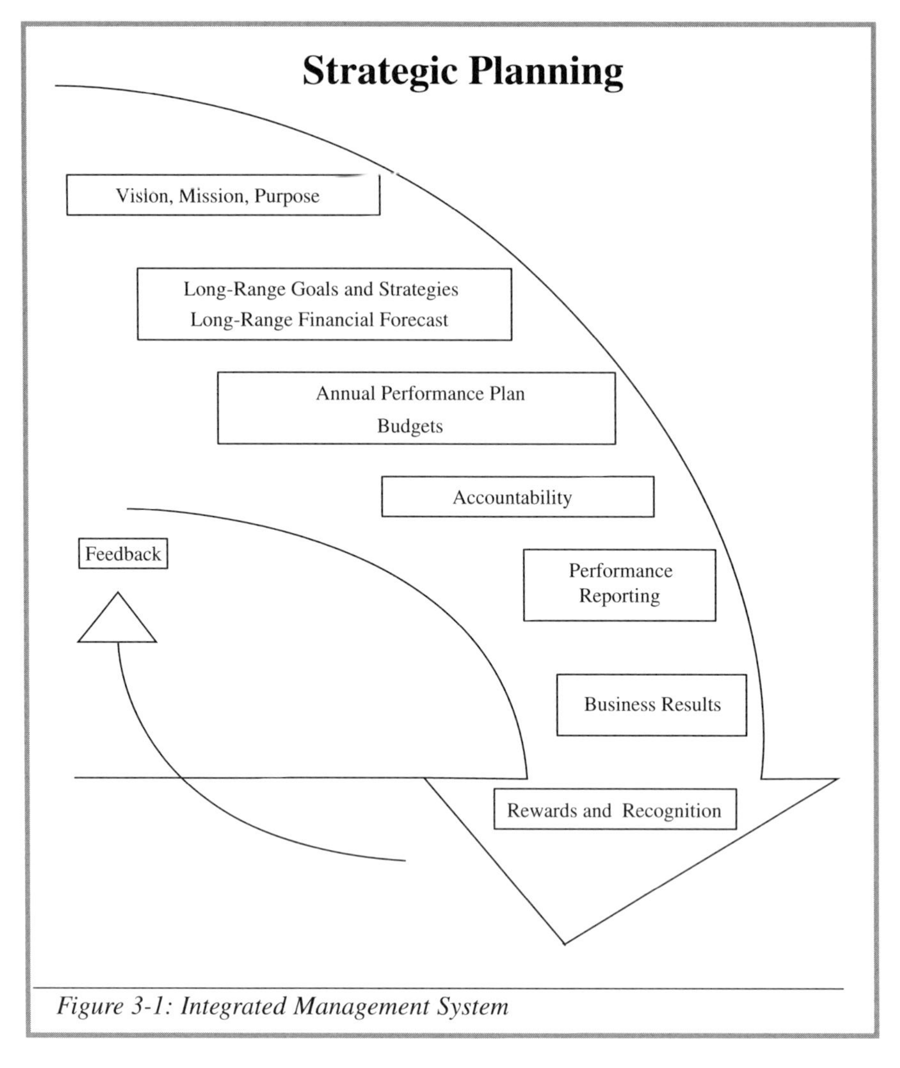

Figure 3-1: Integrated Management System

- *Corporate scope.* The scope defines the corporation's lines of business and geographic areas of operation. As with the corporate purpose, this may be either general or specific. The corporate purpose clarifies exactly what portion of the market the firm plans to occupy to distinguish itself from other firms in its industry. The combined corporate purpose and scope become a concise and clearly stated corporate mission statement.

- *Corporate objectives.* Objectives provide managers with specific goals and are directed toward management accountability. The objectives are stated in quantitative terms, such as target returns, profit margins, and earnings per share, or in qualitative terms such as "maintaining the highest levels of customer service and satisfaction." To be measurable and achievable, objectives are stated in quantitative terms whenever possible.

- *Corporate strategies.* These are the approaches used to realize the objectives within the framework of the corporate purpose and scope. These strategies provide the detailed "blueprints" showing how the managers of the firm will achieve stated objectives. Specifically, strategies address such issues as product slate, choosing markets, pricing, and marketing approach.

- *Corporate directives.* These are the basic set of assumptions concerning overall growth, the economy, inflation, etc. used by different areas of the firm in developing their plans and forecasts. This ensures that all areas will be "in sync" with each other in developing plans.

SCENARIO PLANNING

Royal Dutch/Shell developed scenario planning in the early 1970s. Scenario planning is not used to "predict" the future but as a tool for questioning assumptions to see how the world works. Alternate outcomes are considered beyond a base or platform case. Often in the planning process a preferred illusion of certainty is taken in understanding risks and realities. Eventually, the denial of uncertainty sets the stage for surprises. Scenarios consisting of events and resulting end states consider the driving forces of:

- Social changes affecting demand for petroleum products
- Technological changes resulting in improvement of exploration for oil and productivity
- Economic changes in price of oil, costs of financing and operating expense
- Political changes affecting the continuity and risks of investments
- Environmental laws and their impact on oil and gas operations.

Events, influenced by industry participants, affect the end state, or snapshot of future industry conditions. The sum of the events and resulting end state defines the scenario. End states are purposefully extreme and different. They capture different points of view, designed to explore "what drives what" and are written from the point of view of the planning horizon year. Trigger

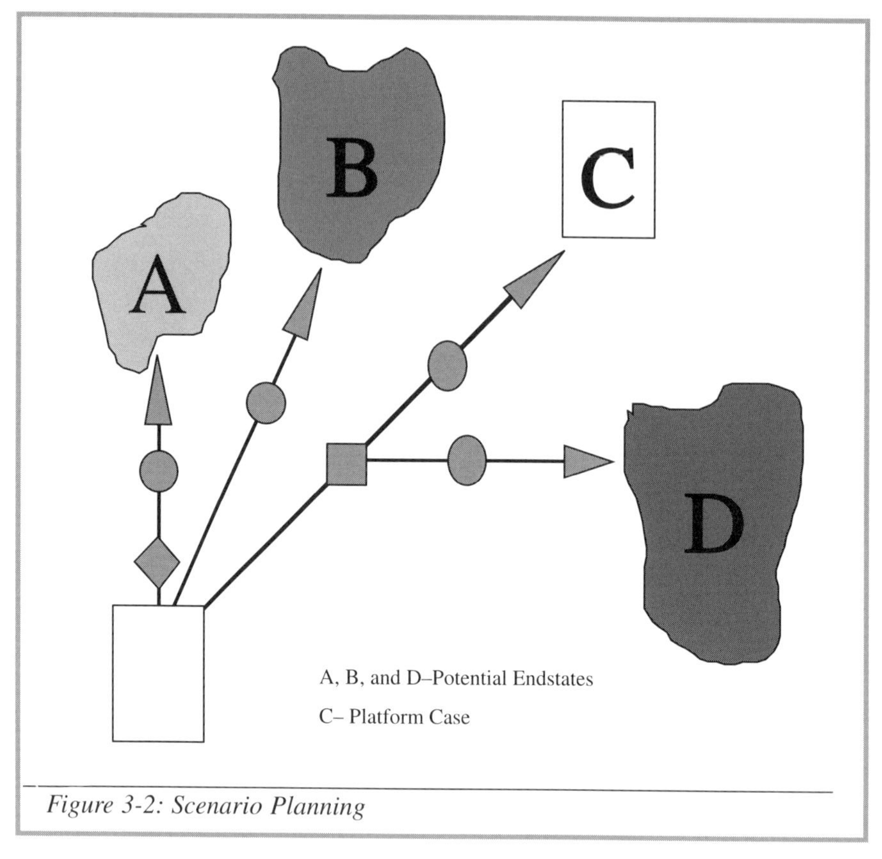

Figure 3-2: Scenario Planning

events affecting the end states are influenced by five forces: power of the suppliers, new entrants, substitutes, competition, and customers.

LONG RANGE FINANCIAL PLANNING

The *long-range financial plan* is an important part of the firm's total operating plan. Its purpose is to use information provided by sales and production forecasts, as well as from capital budgets, to determine what the future financial position and statements of the firm will look like.

The long-range financial plan decides how the company will raise the funds it needs to pay for all the things it wants to do. Cash inflows from operations, as well as projected outflows of cash for the purchase of new assets, dividends to owners and repayment of debt are computed. The sources of funds needed to finance the company are included over the life of the plan.

The plan helps management answer questions such as: How much profit will be generated? How much will be paid in shareholder dividends? What will be the available retained earnings?

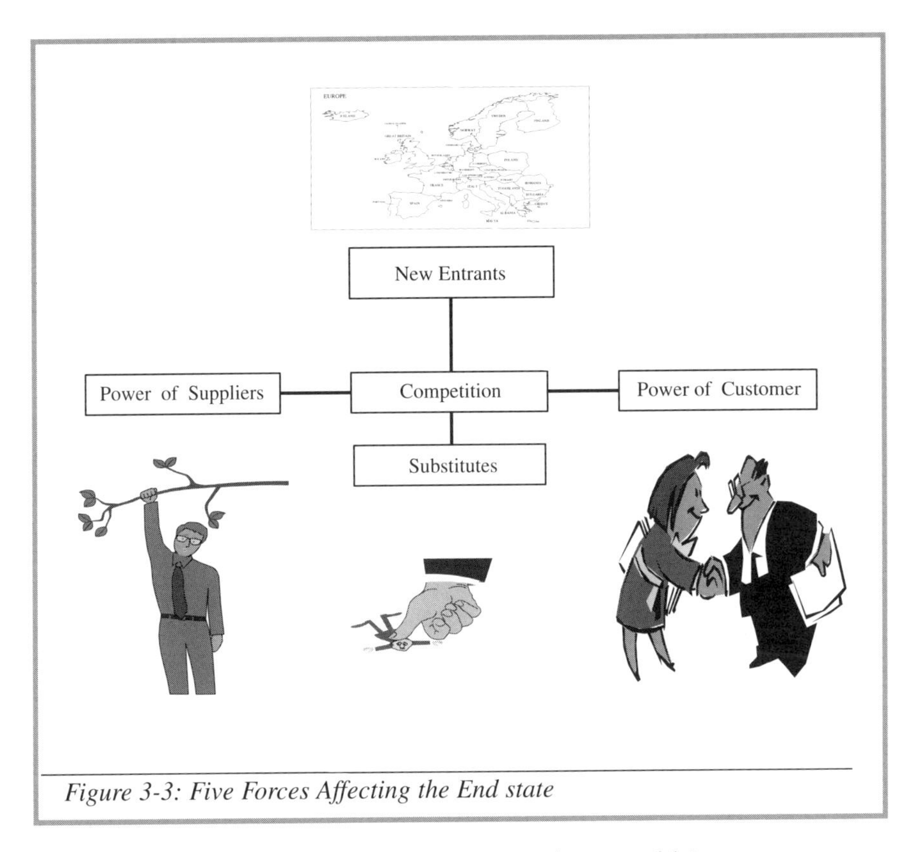

Figure 3-3: Five Forces Affecting the End state

Will new debt be required and should it be short-term or long-term debt?

Forecasting should begin with an evaluation of industry-wide demand. In the energy industry, for example, this begins with an analysis of the general demand for energy products, realizing that alternative energy products may fill some demand. Other aspects of the forecasting process at this "macro" level will include:

- Overall economic/industry trends
- Energy supply/prices (national and international)
- Cost of exploration
- Production capacity
- Refinery capacity
- Refinery throughput
- Transportation capacity
- Capabilities and intentions of competitors

Company Specific Forecast

The next step in the process is the preparation of a company specific forecast. Such "micro" forecasting will include consideration of the following:

- Market concentrations
- Operating/exploration area
- The relative quality of the organization's sales force

Forecasting plays an important role in the planning process because it enables managers to anticipate the future and to *plan* accordingly. Business forecasting pertains to more than predicting cash requirements. Forecasts are also used to predict profits, revenues, costs, interest rates, currency rates, prices and supply of crude, movements of key economic indicators and other variables. However, for the sake of simplicity, the focus of this chapter is on forecasting cash requirements. Nevertheless, the concepts and techniques apply equally well to these other variables.

Features Common to All Forecasts

There are many forecasting techniques in use. In many respects they are quite different—nonetheless, there are certain qualities common to all and it is important to recognize them. These qualities are:

1. Forecasting techniques generally assume that the *same underlying causal systems* existing in the past will exist in the future.
2. Forecasts are *rarely perfect*; actual results will usually differ from predictions and allowances are made for inaccuracies.
3. Forecasts for *groups of items tend to be more accurate* than forecasts for individual items because forecasting errors among items in a group have a canceling effect.
4. Forecast accuracy *decreases* as the period covered by the forecast (i.e., the time *horizon) increases*.

Steps in the Forecasting Process

There are five basic steps in the forecasting process:

1. *Decide the purpose* of the forecast and when it is needed. This provides an indication of the level of detail required in the forecast, the resources (work force, computer, dollars) justified, and the level of accuracy necessary.
2. *Establish a time horizon* that the forecast must cover, keeping in mind that accuracy decreases as the length of the forecast period increases.

"Forecasting"

3. *Select a forecasting technique.*
4. *Gather and analyze the appropriate data*, and then *prepare the forecast*. *Identify any assumptions* made with preparing and using the forecast.
5. *Monitor the forecast* to see if it is performing satisfactorily. If it is not, reexamine the method, assumptions, validity of data, and so on. Modify as needed, and prepare a revised forecast.

There are many different methods used to develop forecasts but we will only discuss and illustrate forecasting based on percentage-of-sales.

FORECASTING FINANCIAL POSITION AND REQUIREMENTS: PERCENTAGE-OF-SALES METHOD

Percentage-of-sales forecasting is a simple yet effective forecasting method that allows managers to quickly estimate the future financial needs of the firm quickly. It is based on two assumptions:

1. Most balance sheet accounts are tied directly to sales
2. The current levels of all assets are optimal for the current sales level (or the existing ratio of assets to sales will continue).

These are critical assumptions and the successful use of this method is highly dependent upon accurate sales/volume forecasts. The steps are detailed below and the financial planning illustration using the percentage-of-sales forecasting method is provided. Steps in developing the percentage-of-sales forecast are:

1. Decide which balance sheet items will vary directly with sales and calculate the percentages of sales for those items for the current year based on actual and historical levels.
2. Use these percentages multiplied by the sales forecast to decide the balance sheet levels for the following year. While depreciation does not seem to be explicitly considered as a source of funds, the current year's depreciation charges are included in the determination of retained earnings.
3. Insert previous levels for balance sheet accounts that cannot be forecasted as a percentage of sales (notes payable, long-term debt and equity). At least one of these will be changed in the final version of the balance sheet.
4. Add the expected increase in retained earnings to the previous year's equity level to obtain the forecast for the equity accounts.
5. Decide the level of external funds needed to balance the accounts; this amount will be added to the various external fund account.
6. Determine the projected allocation of additional funds needed for the debt and equity accounts, adjusting for any restrictions on the firm's debt or liquidity or any other restriction. This allocation should be consistent with the firm's capital structure policy.

Major Features of the Percentage-of-Sales Method

Projected Financial Statements. Given the steps outlined above, the firm can construct a set of pro forma financial statements for the planning period. The pro forma statements are analyzed to decide the financial impact on the firm over the planning horizon.

Analysis of Projected Ratios. Once the projected financial statements are completed, financial ratio analysis can be done to decide the impact of strategies and other assumptions on the future of the firm. These ratios are compared either to past values for the firm or to industry averages or competitors' ratios.

Relationship Between Growth and Financial Requirement. There are relationships between the growth rate in sales and other factors in the forecasting approach that will ultimately affect the general financial requirements of the firm. Those relationships affect the ability of the firm to sustain growth over time without excessive reliance on external sources of financing. In the petroleum industry, major expenditures are for exploration, production, transportation, oil and gas processing, storage facilities, and consumer distribution facilities.

The concept of "sustainable growth" attempts to find a growth rate that the firm can sustain over time without having to raise new capital in the external markets. This means that growth must be "financed" by cash flows from operations and other sources tied directly to sales increases (i.e., increases in accounts payable and accrual items).

Growth and Financial Requirements. At low growth rates, the firm may find that it can generate all the funds it needs through internal sources (retained earnings and spontaneous increases in certain liabilities, such as accounts payable and accrual items). Dividend policy has a major impact on the ability of the firm to grow over time. Essentially, dividends are earnings of the firm diverted to the shareholders rather than reinvested to fund the acquisition of new assets. It is sufficient to understand that most firms will pay some level of dividends and this will have an impact on the need for external funding.

If the firm wishes to sustain a high growth rate, its directors may decide to pay low dividends or none at all, thus increasing the level of retained earnings and reducing the need for external funds. This is often the case for new or start-up firms in high growth industries.

Capital Intensity. As the capital intensity of the firm increases, more assets are required to support a given level of sales. Alternatively, if the firm can increase its productivity (reduce capital intensity) it will also reduce its need for external funding for a given revenue stream. This kind of productivity increase is often seen in firms that are growing and can realize economies of scale in their production processes. The petroleum industry has a high level of capital intensity and future strategies that depend upon capital investments.

Profit Margin. As the firm increases its profit margin, its need for external funding will decreases, all other variables being constant. Again, as a firm grows and begins to realize economies of scale reduce its per-unit costs of production, resulting in increased profit margins. In addition, some costs may be fixed, which offers the firm some operating leverage as the revenue levels increase. The driving force for downstream mergers and consolidations is to create larger and more efficient enterprises. A consolidation normally reduces overhead costs, transportation costs (as the partners reshuffle the logistics that move oil products from refineries to gas stations), and competition that permits consumer prices to increase.

The percentage-of-sales forecasting, method makes some critical assumptions in the development of the *pro forma* financial statements for the planning period. Often these assumptions may not fit a "real-world" context, thus making the model more difficult to use. Fortunately, many planners use computer-based spreadsheets to develop these models, which makes the adjustments for changing assumptions much easier.

Some of the most common problems, together with suggested solutions, are listed below.

Economies of Scale. When these are present, the firm's effective productivity will increase over time as the size of the firm increases. This generally results in a reduction in the ratio of assets to sales, lower per-unit production costs, and higher profit margins due to greater operating leverage from fixed production costs. To adapt to changes in the forecast, the financial planner must forecast new ratios for each of the above items over the planning horizon and use the new ratios rather than constant ratios.

Lumpy Assets. For the oil industry, assets cannot usually be added in small continuing amounts but must be added in large discrete lumps. This is usually the case with fixed assets, where we see a step function in the relationship between increasing fixed assets and sales. In managing this change in the assumptions, the planner must forecast the level of assets (usually only fixed assets) needed for each year of the planning horizon to support the estimated level of revenues.

Cyclical Changes. Unforeseen changes in the economic environment may significantly affect forecast accuracy. As the economy changes, the stream of revenues forecast and/or expenses may change, resulting in an inaccurate set of pro forma statements. The best way to adjust for this problem is to develop several sets of pro forma statements for different economic scenarios. This is especially easy if the forecast is developed in a spreadsheet.

Seasonal Industries. Most of the consumer products in the oil industry have strong seasonality. Firms must either make substantial investments in plant and equipment to produce at a high rate during a short period, or must build up high levels of inventories to meet demand. These factors make forecasting in seasonal industries more difficult.

This factor can be adjusted in much the same way for lumpy assets. The planner must determine the level of assets needed to support the projected sales level during the heaviest production season for each year of the planning horizon. The asset levels are used to develop the forecast.

Strategic Changes. Forecasting becomes difficult if a company changes its strategy by acquiring or divesting a business, or changes its debt structure. The historical financial statement relationships will change. In a major strategic change, the current financial plan may have to be scrapped and a new one developed. If the original plan was broken out by divisions, and if a divestiture occurs, the remaining divisions will be combined to create a new forecast.

An Illustration of Percentage of Sales Method

The percentage of sales method is widely used to forecast a firm's financial requirements and to measure the impact of financial plans and policies. The first step in the percentage of sales method of forecasting is to express various balance sheet items as a percentage of sales. Only asset categories of significant size that vary directly with sales need to be calculated. Cash receivables, and inventories are items that typically increase with an expansion of sales. Fixed assets may not have to increase with sales if the firm is operating at less than full capacity.

Assumptions For Forecasting The Income Statement Using Percentage-of-Sales Method

1. Crude oil purchases and operating expenses are variable and dependent on forecasted sales. Assume 20% increase in sales for Forecast Year 1 and a 10% increase in sales for Forecast Year 2.
2. Depreciation, depletion & amortization, and interest expense remain constant.
3. Taxes are 50%.

	$MM				
	Current-------------------	-Forecast------------------------			
	Year	% Chg	Year 1	%Chg	Year 2
Sales	$ 1,200	20%	$ 1,440	10%	1,584
Cost of Purchased Crude	800	20%	960	10%	1,056
Gross Profit	400		480		528
Operating Expense	162	20%	194	10%	213
DD&A	38	n/c	38	n/c	38
Earnings before Interest	200		248		277
Interest	80	n/c	80	n/c	80
Earnings before Taxes	120		168		197
Taxes (50%)	60		84		98
Earnings after Taxes	60		84		99
Preferred Dividends	18		18		18
Common Stock Dividend	42		66		81
Change in Retained Earnings	0		0		0

Table 3–1 Forecasting Income Statement Using Percentage-of-Sales Method

4. Earnings in excess of preferred stock dividends will be issued as common stock dividends.

Assumptions for Balance Sheet Forecasts Using Percentage-of-Sales Method

Current Assets

1. *Cash* remains constant
2. *Accounts receivable* increases with sales. The new receivables balance is projected by applying the historic *average collection period* to the predicted sales level. Accounts receivables equal 60 days based on sales. Calculate new receivables (sales/365 x 60).
3. *Inventory* will increase with sales. The projected inventory level depends on turn over of inventory based on sales. Inventory turnover is four times yearly based on cost of Crude Purchases. Calculate new inventory (Crude Purchase/4).

Other Current Assets remain constant.

Fixed Assets

1. Changes in fixed assets occur only because of preplanned events reflected in the strategic plan.
2. Depreciation will reduce the balance and can be estimated based on the projected asset and depreciation policy.
3. Fixed assets will decrease annually by $38 million as a charge to depreciation expense and an increase to allowance for depreciation.

Current Liabilities

1. Accounts payable depends on cost of goods sold and based on days' purchases outstanding.
2. Notes payable should not be allowed to increase without a conscious effort reflected in the strategic plan.
3. Accruals are a function of timing and policy and because of the small balance can be estimated based on a percentage of sales.

Long-term Debt/Equity should not be allowed to increase unless specified in the strategic plan.

Equity. Changes in equity vary with the income statement, which identifies the change in retained earnings, and a planned effort to increase or decrease existing equity. The effect of all these projections is a balance sheet that does not balance. Generally, the assets will exceed the liabilities and equity by an amount that we will define as additional or external funding needed and classify as long term liability.

All other accounts remain constant.

$MM	Current Year	% Chg	Forecast Year 1	%Chg	Year 2
Cash	20	n/c	$ 20	n/c	$ 20
Accounts Receivable	200	20% inc	240	10% inc	264
Inventory	200	20% inc	240	10% inc	264
Current Assets	$ 420		$ 500		$ 548
Net Fixed Assets	750	*	712	*	674
Total Assets	$1,170		$1,212		$1,222
Liabilities					
Accounts payable	$ 50	20% inc	$ 60	10% inc	$ 66
Notes payable	200	n/c	200	n/c	200
Accruals	10	n/c	10	n/c	10
Total Current Liabilities	$ 260		$ 270		$ 276
Long- Term Debt	$ 400	n/c	$ 400	n/c	400
Funding Adjustment	0	**	32	**	36
Total Liabilities	$ 660		$ 702		$ 712
Equity					
Preferred Stock (18%)	$ 100		$ 100		$ 100
Common Equity	410		410		410
Total Equity	$ 510		$ 510		$ 510
Total Liabilities & Equity	$ 1,170		$ 1,212		$ 1,222

*$38 Depreciation each year reducing amount of Fixed Assets; ** represents the deficiency of cash under current policy—will need to be borrowed or change dividend policy.Year 2 represents cumulative amount

Table 3–2 Forecasting the Balance Sheet Using Percentage-of-Sales Method

Assumption for Forecasting Cash Flow Statement Using Percentage of Sales Method

- *Cash flow* depends on cash generated by operating and financing functions.
- *Cash from operations* results from profit/loss plus non-cash items plus/minus changes in working capital accounts.

- *Cash from investing* results from changes in plant assets.
- *Cash from financing* results from increases/decreases in equity and long-term liabilities.

	$MM -----FORECAST YEAR------ YEAR 1	 YEAR 2
Cash from Operations		
Net Income	$ 84	$ 99
Noncash Items, Depreciation	38	38
Changes in Working Capital	(70)	(42)
Total cash from operations	$ 52	$ 95
Cash from Investing	0	0
Cash from Financing		
Funding Adjustment	$ 32	$ 4
Dividends	(84)	(99)
Cash from Financing	$(52)	$(95)
Cash Beginning	20	20
Cash Ending	$ 20	$ 20

Table 3–3 Forecasting Cash Flow Using Percentage-of-Sales Method

OTHER APPROACHES TO FORECASTING

There are many forecasting techniques in use. In many respects they are quite different. Nonetheless, there are certain qualities common to all and it is important to recognize them.

Analysis of Time Series is a time ordered sequence of observations that have been taken at regular periods of time. The main theme of time series analysis is that future values of a series can be estimated from past values of the series. There are four commonly used techniques for analysis of time series data.

1. *The naive approach* to forecasting is wonderfully simplistic: The latest observation in a sequence is used as the forecast for the next period. Averaging techniques are used to smooth out fluctuations in a time series.
2. *Simple moving average and weighted moving average.*
   - Using *a simple average approach* involves computing the average value of a time series for a certain number of the most recent periods, and treating that average as the forecast for the future.

- The *weighted average method* is similar to the simple average approach except, unlike the simple moving average that assigns an equal weight to each observation, a weighted average generally weights the most recent observations more heavily than older ones.

3. *Exponential smoothing* is a widely used method of forecasting. Its name is derived from the way weights are assigned to historical data. The most recent values receive most of the weight and weights fall off exponentially as the age of the data increases. The simple exponential method is relatively easy to use and understand. Each new forecast is based on the previous forecast plus a percentage of the difference between that forecast and the actual value of the series at that point.
4. *Regression analysis* is often useful where there is a general trend in the relationship between sales and the variable in question. The objective in linear regression is to obtain an equation of a straight line that reduces the sum of squared vertical deviations of points around the line. Computerized financial planning models are available for regression analysis, percentage of sales relationships, and time series analysis.

ANNUAL PERFORMANCE PLANS

Annual performance plans are developed based on, and in support of, strategic plans. As firms delegate more authority, the traditional budget process has been modified to an annual performance plan. The plan defines business performance in terms of milestones and targets (strategic, operational, and financial) to be met during the year.

The Strategic Implementation model builds its budgets for each operating unit to best carry out the overall corporate strategy. Because the focus of the budget process is the mission statement, it is important that department managers give considerable thought to the mission and how they will monitor the department's annual progress toward accomplishing the mission. A clear and concise mission statement with measurable results is the department manager's first step.

An understanding of effectiveness and efficiency is required for the next step. *Effectiveness* is defined as "the accomplishment of a desired result or the fulfillment of a purpose." Therefore, an effective budget is one that accomplishes the mission by achieving the desired results within the required period. However, focusing too much on effectiveness can cause problems. A budget may be effective in accomplishing the mission, but it might do so at the expense of the organization. For example, a marketing department's mission may be to increase market share from 40% to 45%. To achieve this goal, the approach may be to budget for 40 people—a doubling of sales force. While this action may be very effective, it is obviously costly. Therefore, the budget process should be designed to ask the question, "Can the department achieve the same results with fewer people, or is an entirely different approach needed?"

Efficiency, as used in the budget process, means economic efficiency measured by units of output per unit of input. Efficiency increases as more output is produced per unit of input. Therefore, for a budget to be effective and efficient it must identify the resources needed to accomplish the organization's mission and also identify the minimum quantities of those resources.

The traditional budget tries to establish spending levels (limits) based on projected revenues derived from a strategy developed independently by the company's highest levels. The Strategic Implementation Model builds its budgets for each operating unit to best implement the overall corporate strategy.

When re-engineering the budget from a traditional approach to a Strategic Implementation Model, do the following:

- Give up the traditional accounting model, where every general ledger account number on the income statement has a budget amount. Instead, use larger control groups, which give some freedom to the manager to make decisions about spending.
- Go beyond revenues and costs to focus broadly on investment financing, productivity, rates of return, trends, and measures of strength.
- Substitute a uniform performance model that will be applied to all units for the old uniform method of allocating resources among smaller and smaller units.
- Analyze prior and current period financial performance in terms of certain key performance areas and key results areas. Evaluate the relationships between performance and results to reveal the problems and opportunities for improvement.
- Determine which key performance areas and key results areas, if used consistently, could guide mini-strategic plans and how they might affect communication between management groups.
- Consider changing from a "created at the top and imposed down the line" budgeting process. Substitute a more interactive process so that budgets are put together at all levels from both directions-bottom up as well as top down. Organizations vary greatly as to complexity and the degree of management autonomy. In some cases, corporate finance staff can formulate the mini-strategic plans. In other cases, the operating units can and should do it.
- Explore the use of powerful new computer tools to raise the budget process to a par with advances in information technology.

BUDGET REVISIONS

An additional consideration in the operation budget is the company's policy with respect to budget revisions. It is important that there be some mechanism to allow revision of the budget when unanticipated internal or external events occur.

Inflexible budgets in such circumstances reduce the effectiveness of the budget as a control device because it will be perceived throughout the organization that the budget is no longer realistic. However, to the contrary, there is also a danger in allowing the budget to be revised too readily. If this is allowed, the final budget, in the most extreme case, will simply mirror actual results. This will also lessen the effectiveness of the operating budget as a control device, since managers will know that the budget can easily be revised.

Once a company decides how it will deal with budget revisions, it is necessary that such revisions be prepared with as much care as used in the preparation of the original budget. It is also necessary that revisions be carefully coordinated, analyzed, and communicated to all levels of management as when the original budget was prepared.

INTERNAL PERFORMANCE REPORTING

Performance reporting is an ongoing process for monitoring and reporting progress to ensure that agreed-upon actions are executed and targeted results are attained. The milestones and targets established during strategic planning and annual performance plans are the basis for balanced scorecard reports.

The reports are prepared to meet the specific needs of managers at different organizational levels. The degree of emphasis on strategic vs. operational vs. financial performance will likely vary by management level. However, some standardization will be necessary to achieve a common corporate view and to provide consistent communication across the organization.

Balanced reporting should be limited to critical measures, thereby eliminating unnecessary activity and analysis. Report content is decided by the information needs of the recipient. Variance analysis is limited to a few key items. In addition, performance reports provide feedback used to support employee performance goals and reward system and data for the strategic planning process.

Finally, several key characteristics of performance reporting are:

- Feedback from performance reports should drive necessary mid-course corrections to assure attainment of the agreed upon goals.

- Performance reports are issued each quarter and report progress for each operating company and a select number of businesses. The frequency of reporting to other levels of management and the frequency of reporting certain milestones may vary depending upon management's information needs.
- Progress discussions are conducted as needed. The corporate, operating, and specific business units determine which issues call for discussion and debate. This differs from the past in which each business presented an entire update of performance during management meetings.
- There should be frequent, informal, and ongoing communication throughout the organization in addition to the formal performance reporting described above.

"Internal Reporting"

Making The Information Effective

It may seem unnecessary to discuss at any length the importance of statement or report presentation. Yet costs will not be controlled, sales efforts will not be directed into the proper channels, and profit planning will not be effective unless facts are presented to executives and supervisors in such a manner that they can understand them and take action. Merely to present the facts is not enough. These facts must be understood; their significance must be realized by the management team; and management must be motivated to take action.

Types Of Reports

Repetitive reporting is a type of report issued on a regular schedule. It should be in a specially developed consistent format. Its purpose is well known to the recipients as is the nature and meaning of the information being presented. Examples of repetitive reports are the monthly statements of income and expense and financial position and the daily or weekly reports of sales, cash, and manufacturing costs.

Special reports are those prepared to identify a specific issue-for example, a cost item growing faster than planned-or to make a major capital investment decision. Unlike repetitive reports, the format of a special report is whatever is pertinent to the contents. The outline of a special report should include: 1) introduction in which the issue is stated, 2) the analytical approach, 3) highlights or exhibits, 4) analysis and conclusion with implication, and 5) alternatives and recommendations.

Five Basic Principles of Report Preparation

1. The responsibility concept should be employed in that dissemination of facts and figures should relate to the segment of the organization being reported on. The reporting should follow the organization chart in that each manager receives a detailed report regarding his function with the data summarized to the next reporting level.
2. The exception principle should be applied as much as possible. Generally speaking, for control purposes, the out-of-the-ordinary operations should be emphasized. Normal or routine situations do not receive prominence.
3. Figures should be comparative. Actual performance data alone usually are of little significance. Rather, actual performance must be compared with a target or reasonable yardstick. Comparison with budget, standards or past performance is necessary.
4. Data should increasingly be in summary form for each successively higher level of management.
5. Reports should include interpretative commentary or be self-explanatory.

Forms Of Accounting Reports

1. *Written* – tabular, expository, graphic.
2. *Oral* – with visual aids.
3. *Electronic* – report delivered via a computer network.

Graphics

1. *Line* or *curve charts* used for trends.
2. *Bar charts* contrasting quantities over a period of time.
3. *Area diagrams,* or pie charts, comparing parts of a whole.
4. *Solid diagrams* consisting of various geometric forms to illustrate comparison of magnitudes. Accuracy not provided.
5. *Maps* presented in pictorial form, facts in a geographic distribution.

When timeliness is critical in the usefulness of financial figures, informal reports are desirable. Monthly operating results in summary form to meet the needs of top management can be handwritten or electronic reports. Timeliness and communication of critical data is the goal.

Oral communication is an important phase of the reporting function. Providing information is the primary goal of management accountants and oral communications skills should be developed early in their training. A successful oral report will include the following attributes:

1. In the early planning stages, establish two or three issues the audience must leave the meeting with. Do not complicate with a multitude of facts and figures. Limit the number of points. Don't try to cover the whole "waterfront".
2. Visual aids are critical and must be neat and easy to read. Limit word slides to eight words across and eight lines down.
3. Presenter should have a professional appearance, be comfortable and relaxed but enthusiastic. Know the material and encourage questions.
4. As a general rule, do not read the script. If reading, maintain eye contact with audience. Have a logical sequence, be as brief as possible while covering all relevant points.

SUMMARY

The integrated management system is a coordinated effort toward achieving the goals of the organization. The strategic planning process looks into the future and creates a road map for the organization in accomplishing its objectives. The long-range financial forecast is an effort to determine future financial needs of the organization and examine if the strategic plans will result in increasing shareholders' value. The annual budget provides a short-term focus to determine that short-term results contribute to achieving the overall long-range strategic goals. The reporting process is an effort to motivate performance directed toward achieving the overall goals of the organization.

Competitive Comparison

4

Performance evaluation is a critical management tool. It provides perspective on the position the organization holds in the industry. Furthermore, in order to set realistic goals, management must know what the industry standards are. Comparing a company with its peers is a reliable method for self-evaluation and gauging performance. The key to this analysis is the composition of the peer group and the timeliness and accuracy of the data. Peer group determination is usually based on size measured in terms of revenues, total assets and business focus.

There are a variety of measures, such as key financial ratios (listed at the end of this chapter), which normally focus upon efficiency and profitability that can be used throughout the business environment. But evaluation of economic performance goes beyond the financial statements. Reliance on the standard ratios to measure profitability, and efficiency of exploration and production (E&P) activities may be grossly inadequate and incomplete. The deficiencies in standard ratios result from the unique characteristics of upstream activities. They result from the importance to the E&P company of reserves of oil and gas in the ground, the high level of risk involved, and the emergence of alternative accounting principles for oil and gas producing companies.

"Peer Group Analysis"

FINDING COSTS

It is widely understood that the impact of a discovery on the balance sheet or on reported income will not be seen immediately. This is true of both Successful Efforts and Full Cost accounting. Similarly, the result of a dry exploration well may or may not show up in the accounting period in which it occurred, depending upon the accounting convention used. The cost of finding oil and gas is conceptually a simple matter of calculating the cost to find hydrocarbons divided by the quantity of oil and gas found. In practice the matter is not so simple because there are numerous variables to consider such as:

- Which reserve estimates should be used?
  - Discoveries
  - Acquisitions
  - Revisions
  - Enhanced Recovery
- Which costs should be allocated?
- Which accounting method–Full Cost or Successful Efforts?
- How should reserve revisions be treated?
- What time period should be used?
- Which conversion factor should be used to convert gas to oil or vice versa?
- Which regions, if any, should be reported separately?

Finding costs are ordinarily defined as exploration and appraisal expenditures divided by volumes added from extensions and discoveries. In any given year this statistic can lack for real meaning, but with the combined results of over, say, a five-year period, worthwhile trends can be seen. The accuracy of this statistic depends upon both the accountants and their assessment/allocation of costs as well as the reservoir engineers and their estimate of proved recoverable reserves.

Unfortunately, there are few standards for reporting finding costs. Thus, the use of a finding cost should be accompanied by additional information, particularly regarding how the statistic was calculated. The typical early life cycle associated with finding costs and various other measures of quantity and value of reserves added to the books can span a number of years. A typical life cycle of a large oil field is shown in Table 4–1.

Activity	Date
Initial exploration costs expended	1992
Additional exploration costs expended (deliniation/appraisal)	1993
Officially discovered	1994
Bulk of exploration costs reported	1994
Reserves booked	1996
Development costs incurred	1997+
Reserve revisions add reserves	2004
Reserve revisions add more reserves	2008
Enhanced recovery begins	2010

Table 4–1 Life Cycle of a Large Oil Field

The reserves initially booked prior to development are some of the most inaccurate and uncertain estimates of reserves that exist. Yet these reserves estimates are more closely associated in time with the costs associated with a discovery.

Finding costs have wide usage and are often quoted. Like many ratios used to indicate company performance, a single year statistic is not very meaningful. Three- or five-year moving averages are better and tend to normalize the erratic nature of the rate of discovery of oil and gas for a company. Some of the approaches are summarized as:

$$A.\ \textit{Finding costs} = \frac{\textit{Exploration expense only}}{\textit{Reserve additions (excluding revisions)}}$$

$$B.\ \textit{Finding costs} = \frac{\textit{Exploration expense only}}{\textit{Reserve additions (including revisions)}}$$

$$C.\ \textit{Finding costs}^{*} = \frac{\textit{Exploration and development expense}}{\textit{Reserve additions (including revisions + enhanced recovery)}}$$

*Better known as Finding and Development Costs

Sometimes acquisition costs are also included and sometimes not. This is part of the problem associated with the lack of standardization. Results of these three different approaches are shown in Table 4–2.

	A $/BOE	B $/BOE	C $/BOE
Amoco	6.59	3.20	6.15
ARCO	4.19	2.93	4.46
Chevron	8.75	2.90	5.23
Conoco	5.58	3.27	6.69
Exxon	4.53	3.18	9.27
Marathon	2.68	2.38	5.68
Mobil	5.97	2.34	3.54
Phillips	3.53	1.61	2.74
Texaco	4.71	1.81	4.04
Unocal	2.04	1.93	4.43
Simple Average	$4.86	$2.56	$5.22

From: Gaddis, D., Brock, H., Boynton, C. Oil & Gas Journal, 1 June, 1992, pp. 93–95
(Data prepared by Jeff Boone.)

Table 4–2 Worldwide Finding Costs Based on Three Approaches. Five-Year Average (1986–1990).

	3-Year Average (1988–90) $/BOE	5-Year Average (1986–90) $/BOE	5-Year Average (1989–93) $/BOE
Amoco	6.07	6.15	5.96
ARCO	5.42	4.46	5.47
Chevron	5.30	5.23	5.48
Conoco	7.82	6.69	7.71
Exxon	6.36	9.27	6.20
Marathon	5.02	5.68	6.04
Mobil	3.64	3.54	3.85
Phillips	3.05	2.74	3.14
Texaco	4.13	4.04	4.37
Unocal	4.25	4.43	4.57
Simple Average	$5.11	$5.22	$5.28

Table 4–3 Finding and Development Costs (Approach C).

Finding and Development Costs

Some companies report the cost of finding and developing reserves (sometimes also referred to as replacement costs or finding and acquisition costs including revisions). The biggest problem with finding costs is that there is no indication of the value of the reserves found. To illustrate this point, an example is shown in Table 4–4 of two similar discoveries. In each case the finding costs associated with the reserves found amount to $200 million.

For the sake of simplicity, it was assumed that capital costs associated with the development of each field would be the same. This might not be the case, and of course, operating costs are important too. However, the division of profit is the variable that usually makes the biggest difference. This would be a function of the terms of the contracts or fiscal terms in each country. This is discussed further in Chapter 7.

	Malaysia	Australia
Discovery size (million Barrels) (Gross to 100% working interest)	100	100
Estimated Development Costs ($/BBL)	$3.50	$3.50
State Take	88%	45%
Value of the discovery NPV 10%	$48 MM	$240 MM
Entitlement (MMBBLS) (Reserves "Booked")	45	100
Reserve Value ($/BBL) (Prior to Development)		
Working Interest Barrels	$0.48	$2.40
Entitlement Barrels	$1.07	$2.40
Finding Costs ($/BBL)		
Working Interest Barrels	$2.00	$2.00
Entitlement Barrels	$4.44	$2.00
Value Added Ratio		
Working Interest Barrels	.24	1.20

Table 4–4 Value of Reserves Comparison.

Value of Total Proven Reserve Additions

This is a measure of the quality of reserves added. The ratio is the sum of the value added from the yearly changes in standard measure of oil and gas (SMOG) divided by barrels of oil equivalent added.

	Extensions Discoveries & Improved Recovery	Revisions	Total Value
Conoco	3.75	5.73	4.28
Amoco	3.71	2.51	3.18
OXY	3.49	4.20	3.51
Chevron	3.37	7.71	5.07
Phillips	3.31	2.21	3.01
Texaco	3.19	5.02	3.91
Exxon	3.16	2.15	2.70
Mobil	2.46	8.76	5.07
Simple Average	$3.38	$4.79	$3.84

Table 4–5 Value of Total Proven Reserve Additions. SEC 10 Value/Barrel Oil Equivalents (BOE) 1989-1993

Value Added Ratios

Some of the more valuable analytical techniques have evolved from the FASB 69 reserve disclosures. The concept of the value added ratio (VAR) evaluates changes in the standard measure of future net cash flows relative to finding and development costs. Regardless of the

	1989	1990	1991	1992	1993	Average
Conoco	1.75	1.22	1.14	1.16	1.56	1.37
Amoco	1.14	1.44	1.09	0.87	1.25	1.16
OXY	1.74	2.03	1.56	1.82	2.50	1.93
Chevron *	1.97	1.42	1.59	1.82	1.86	1.73
Phillips	1.78	3.12	1.20	1.68	1.30	1.82
Texaco	2.24	2.51	1.56	1.73	1.78	1.96
Exxon **	0.63	1.16	1.42	1.23	1.18	1.12
Mobil **	2.34	2.87	1.94	1.65	1.79	2.12
Unocal	1.90	1.58	1.38	1.41	0.98	1.45
Marathon	0.93	2.14	1.25	1.14	1.32	1.36
Arco	1.31	1.63	0.89	1.49	0.83	1.23
Average	1.61	1.92	1.37	1.45	1.49	1.57

* Does not include interests in Kazakstan–Tengiz
**Exxon and Mobil included acquisition costs of proved reserves.

Table 4–6 Pre-Tax Value Added Ratio, 1989–1993

corporate cost of capital, the 10% discount rate must be used because this is the statutory discount rate for the standard measure reserves disclosure under the SEC guidelines. The value of the reserves is often referred to as the "SEC 10 value." From one year to the next, the SEC 10 value should increase and this is compared to the finding and acquisition costs incurred during that period. The VAR approach compares changes in standard measure with exploration costs incurred.

	Pre-Tax SEC10 Value $/BBL	Finding Costs $/BBL	Pre-Tax Value Added Ratio
Conoco	4.28	3.12	1.37
Amoco	3.18	2.74	1.16
OXY	3.51	1.82	1.93
Chevron *	5.07	2.93	1.73
Phillips	3.01	1.65	1.82
Texaco	3.91	1.99	1.96
Exxon **	2.70	2.41	1.12
Mobil **	5.07	2.39	2.12
Unocal	2.96	2.04	1.45
Marathon	3.64	2.68	1.36
ARCO	5.15	4.19	1.23
Simple Average	3.86	2.54	1.57

*Does not include interests in Kazakstan–Tengiz

**Exxon and Mobil included acquisition costs of proved reserves

Table 4–7 Pre-Tax Value Added Ratio, 1989–1993

Pre-Tax Value Added Ratio = Pre-Tax SEC 10 Value ÷ Finding Costs

Table 4–7 reports the running averages of pre-tax SEC 10 value, finding cost, and pre-tax value added ratio. In comparing the results of the different companies, consideration should be given to where the exploration and drilling occur—terrain, offshore vs. onshore, geographic location, etc.

Comparison of reserve replacement is made in Table 4–8 between geographical location, type of company and type of additions.

	Additions Only	Additions and Revisions	All Sources
United States			
Majors	$5.46	$4.14	$4.05
Independents	5.79	5.41	4.85
Weighted Avg.	5.60	4.61	4.44
Foreign			
Majors	7.10	5.29	5.07
Independents	6.94	5.38	5.00
Weighted Avg.	7.08	5.30	5.06

From Arthur Anderson Oil & Gas Reserves Disclosure 1996

Table 4–8 Reserve Replacement Costs, $/BOE 1993–1995

	Additions Only	Additions and Revisions	All Sources
United States			
Majors	61 %	77 %	68 %
Independents	97	114	157
Weighted Avg.	67	83	84
Foreign			
Majors	75	111	114
Independents	100	134	128
Weighted Avg.	78	114	115

From: Arthur Anderson Oil & Gas Reserve Disclosures 1996

Table 4–9 Oil Production Replacement Rates (%), 1993–1995

Reserve Replacement Rates

It is clear for the petroleum industry that reserves must be found to replace production. Otherwise, a company is simply liquidating its primary assets. Furthermore, a company must find more oil and gas than it produces or it will not grow.

	Additions Only	Additions and Revisions	All Sources
United States			
Majors	67 %	94 %	92 %
Independents	113	117	154
Weighted Avg.	86	103	118
Foreign			
Majors	97	111	99
Independents	109	140	114
Weighted Avg.	99	115	101

From Arthur Anderson Oil & Gas Reserve Disclosures 1996

Table 4–10 Gas Production Replacement Rates (%), 1993–1995

Market Value Added

Market value added (MVA) approaches measure the growth or shrinkage of shareholder wealth. MVA is the difference between what shareholders have contributed to the firm and the market value of the firm's capital stock.

Some definitions focus exclusively on equity contributions and stock values, which ignore leveraging. Other formulas compare the total capitalization of a corporation, including debt and equity, with the corporation's market value including equity and debt. MVA and the related economic value added (EVA) have become popular analytical techniques. Stern Stewart & Company coined the terms.

The equations for MVA can range from the simple to the complex. Three examples are shown:

1. MVA = Market value of equity – Equity capital provided

Example Calculation

	Share Price	*# of Shares*	
Market Value	*$36.0*	*2.7 MM*	*$97.2 MM Equity Market Value*
Book Value	*$22.5*	*2.7 MM*	*$60.7 MM Book Value*
Market Value Added			*$36.5 MM*

*2. MVA = (Market value of equity + market value of preferred stock + market value of debt) – total capital**

Definition 2 is more appropriate. This calculation of MVA requires identifying all the capital a company has taken in including equity, debt, bank loans and retained earnings less the firm's total market value, which includes market capitalization as well as market value of debt.

*3. MVA = (Market value of equity + preferred stock + debt + dividends) – total capital**

*Total capital = Market capitalization + market value of debt and preferred stock

Some adjustments may be made for aspects of capital employed that are not represented on the balance sheet. The common example is research and development costs that have been expensed through time. Techniques are employed to capitalize these elements and include them. These adjustments are an effort to convert the corporation's accounting book value to an economic book value which is a better measure of the cash that investors have contributed. These are referred to as equity equivalent (EE) adjustments.

Assume that over the past five years, the firm has spent and written off $17 million in research and development (R&D). It might be likely that the economic value of the R&D expenditures may last well into the future. A decision is made to capitalize these expenditures and add this to the book value of capital employed called an '*equity equivalent* (EE).' In the formula below, this adds $5 per share. It also provides a better comparison of value created for shareholders vs. the total of their contributions.

Calculating MVA with EE adjustment

	Share Price	*# of Shares*	
Market Value	*$36.0*	*2.7 MM*	*$97.2 MM Equity Market Value*
Book Value + EE	*$27.5*	*2.7 MM*	*$74.3 MM Book Value Adjusted for EE*
Market Value Added			*$22.9 MM*

An objection that might be raised to capitalizing R&D is that it may leave assets on the company's books that no longer have value. What if the R&D fails to pay off? Shouldn't at least the unsuccessful R&D outlay be expensed? The answer is no. Full-cost accounting is the only proper way to assess a company's rate of return.

This argument also surfaces in the context of expensed unsuccessful exploration drilling. This is typical of the Successful Efforts vs. Full Cost accounting issues. With Successful Efforts accounting, a company capitalizes only the costs associated with actually finding oil; all exploration drilling expenditures that fail to discover economic quantities of oil are immediately expensed. This kind of policy reduces earnings early on relative to Full Cost accounting results, but it causes a permanent reduction in assets that eventually leads to overstatement of future rates of return. The same is true of MVA. With Full Cost accounting, by contrast, an oil company capitalizes all drilling outlays including unsuccessful exploration efforts. Experience in the oil industry indicates that part of the cost associated with finding oil is that unsuccessful wells have to be drilled. Just as R&D outlays should be capitalized and amortized over their projected lives, not because they always do create value, but because they are expected to. Therefore, Full Cost accounting must be employed either directly or with EE adjustments to properly measure an oil company's capital investment and thus its true rate of return or MVA.

The accounting model relies on two distinct financial statements: an income statement and a balance sheet, whereas the economic model uses only one—sources and uses of cash. Because earnings are emphasized in the accounting model, it makes a great deal of difference whether a cash outlay is expensed on the income statement or is capitalized on the balance sheet. In the economic model where cash outlays are recorded makes no difference at all, unless it affects taxes.

Thus, at the very least, the MVA approach would require adjustments in order to place Full Cost and Successful Efforts companies on equal footing for comparison. Furthermore, two firms could have similar results; but if there is a difference between them regarding dividends paid, then the results can be misleading. For example, if a company pays out a large dividend, it might not look as healthy from a MVA point of view as compared to a company that has not paid dividends.

Economic Value Added (EVA)

Economic Value Added measures the creation of wealth (value) in a given accounting period or series of accounting periods. This approach requires a perspective on the cash flow generating capability of the firm compared to the firm's cost of capital. The standard weighted average cost of capital (WACC) approach is used.

EVA = Discretionary cash flow – cost of capital
or
EVA = Operating profit after tax – cost of capital

Calculating EVA

After-tax operating profit	=	*$12.0 million*
Cost of capital of 11% x Total capital of $80 million	–	*8.8 million*
Economic value added (EVA)	=	*$ 3.2 million*

Like many of these concepts, there are various means by which they can be calculated.

BENCHMARKING

Benchmarking a company's financial results with other companies has been done for years. However, in the late 80s and early 90s another type of benchmarking was formalized and has become popular. In most companies, the focus on improvement has resulted in reorganizing and business process improvement. After analysis of business processes is performed within the company, it is at this point that people usually ask, "Is it possible to improve the process? If so, what can we do to make it better?" This requires a comparative analysis of similar processes within the organization and outside the organization as well. In fact it is advisable to look outside the industry for fresh perspective.

Benchmarking is a means by which companies can jointly perform a comparative assessment of routine operations in an effort to find ways to economize and enhance efficiency. Companies will not disclose proprietary secrets or exploration strategies, but there are areas where comparing notes can be beneficial. Examples include staffing platforms, helicopter/transportation rates, inventory management, automation, catering, and supply boats. The focus on most benchmarking surveys is operating practices of existing wells, platforms, and gas plants, etc.

The purpose is to understand what others are doing and to use this combined experience and knowledge to help develop improved processes or practices. *This act of systematically defining the best systems, processes, procedures, and practices is called benchmarking.*

Two primary reasons for using benchmarking are for goal setting and process development. Every person, process and organization needs goals to strive for. Even more importantly, benchmarking provides a way to discover and understand methods that can be applied to effect major improvements. It not only shows what realistic goals may exist, but how to get there.

Benchmarking involves:

- Deciding what will be benchmarked
- Defining the processes to compare
- Developing measurements to compare
- Defining internal areas and external companies to benchmark
- Collecting and analyzing data
- Determining the gap between existing and best processes or practices
- Developing action plans, targets, and measurement processes
- Updating the benchmarking effort

Benchmarking should be a continuous discovery activity. As soon as data is no longer added to the database, it becomes outdated. Public domain data should be added regularly to the database, and someone should be assigned to review these new inputs to identify specific breakthroughs. Every one to three years, the total benchmarking effort should be repeated. This is absolutely necessary with the changes that are occurring so rapidly.

Benchmarking Partners

Benchmarking has become an increasingly popular learning exercise for finance organizations bent on change. Once banned, the idea of sharing information on costs and processes with other companies, even competitors, is now being embraced. The concept is quite simple: Find out who is the best in an area you want to improve on and see how they do it.

The Hackett Group, a management consulting firm in Hudson, Ohio, that specializes in corporate finance, has compiled one of the most extensive benchmarking databases (with some 70 companies participating). In April 1993, the American Institute of Certified Public Accountants developed a joint venture in which it will expand Hackett's database and make best-practices information available to its members.

Another database has been developed by the newly formed Continuous Improvement Center (CIC) of the Institute of Management Accountants in Montvale, New Jersey, in conjunction with Gunn Howell Markos Partners Inc., a New York-based consulting firm. The CIC began accepting members in July 1993.

Arthur Andersen & Co., in a joint effort with member companies of the Houston-based American Productivity & Quality Center's International Benchmarking Clearinghouse, has developed a process classification system aimed at standardizing performance and process analysis. In addition, companies within specific industries have formed databases for comparing operational as well as accounting processes. Ernst and Young has benchmarking efforts for pipelines. Solomon has comparative databases for refineries.

Benchmarking is not a new concept. It has been used for years to study competitive financial results and products. The refinement of techniques in recent years has led to its extensive use in defining the best business practices. Systematic benchmarking can elevate a business process to the highest level of efficiency and effectiveness. Without benchmarking, organizations may never truly know how good they are, how good they should be, or what they might strive for.

KEY FINANCIAL RATIOS

This chapter has focused on indices considered to be important and unique in the oil and gas industry. However, this chapter would not be complete without mentioning the traditional financial ratios used in comparing the position of a company from a strictly generic financial analysis. Listed below are some of the more common financial ratios used in analyzing financial position.

Coverage Ratios

Cash Flow Interest Coverage = *(Cash Flow + Provision for Income Taxes + Interest Expense – Deferred Taxes) ÷ (Interest Expense + Preferred Dividends)*

EBIT Coverage = *Earnings Before Interest and Taxes ÷ Interest*
= *(Earnings + Provision for Income Taxes + Interest Expense) ÷ Interest Expense*

Leverage Ratios

Current Ratio = *Current Assets ÷ Current Liabilities*
Quick Ratio = *Liquid Current Assets ÷ Current Liabilities*
All Long-term Obligations = *Long-term Debt + Preferred Stock + Other Long-Term Liabilities (does not include current liabilities)*
Interest Rate = *(Annual Interest Expense + Preferred Dividends) ÷ (All Long-Term Obligations)*
Working Capital = *Current Assets – Current Liabilities*
Percent Debt = *All Long-Term Obligations + Current Liabilities ÷ Total Capitalization*

Performance Ratios

P/E Ratio	*= Market Value Per Share ÷ Adjusted Earnings Per Share*
P/CF Ratio	*= Market Value Per Share ÷ Cash Flow Per Share*
Debt Adjusted P/CF	*= (Market Capitalization + Long-Term Debt + Other Obligations) ÷ (Cash Flow + Interest Expense) (Interest expense can be "tax adjusted" by applying either statutory or effective tax rates.)*
Debt Adjusted Cash Flow Multiple	*= (Stock Price/Share + Debt/Share) ÷ Cash Flow/Share*
Market Value/Share	*= Trading value of common stock as of the effective date of analysis*
Return on Assets	*= Adjusted Net Income ÷ Reported Total Assets*
Return on Equity	*= Adjusted Net Income ÷ Book Value*

Cost Management in the Oil and Gas Industry

5

Cost pressures in the oil industry are likely to remain for many years due to the chronic overcapacity and increasing competition of new global players. Most oil companies believe that oil prices will remain flat for the next few years and are looking at ways to improve and better manage their cost structure.

Management's focus has changed from simply determining the cost of products or service to supporting cost management systems. This chapter begins with an overview of a total cost-management system that identifies how management's decisions affect costs. To do so, it first measures the resources used in performing the organization's activities and then assesses the effects of potential changes in those activities. Potential changes in the activities are then eliminated, improved or automated. Today's technology is helping the "change to" activities and opening doors for more challenging and value adding opportunities.

TOTAL COST MANAGEMENT

Total Cost Management (TCM) is a business philosophy of managing all company resources and the activities that consume those resources. Managing costs means focusing on the factors that cause or "drive" cost consuming activities and developing cost management strategies to achieve cost objectives. (See Table 5–1).

In addition, the focus on international trade requires a common definition of quality and competitive costs. The quality movement is reinforced by the United States-based Malcolm Baldridge National Quality Award and the Geneva, Switzerland-based International Organization for Standardization (ISO). These organizations have developed procedures for measuring quality in an organization to ensure international standards.

Basic to quality measurement is the idea that all goods or services from an organization are a result of multiple processes. All processes have inputs from external or internal suppliers, a value adding transformation involving people, and outputs. Output will consist of products or services for internal or external customers. Analysis of the processes begins with focusing the

• **Recognizing the importance of segment profitability**, because deep spot markets now permit the separation of the upstream from the downstream. Some companies are even separating refining from marketing.
• **Divesting unprofitable and nonessential assets**. Companies have left such noncore business as minerals, office products, department stores, and hotels. They are also selling or swapping poorly performing oil and gas assets.
• **Reallocating investments toward profitable projects**. Companies are shifting their exploration and production funds around the world to improve profitability.
• **Reducing the scope of operations**. Companies are concentrating on more profitable geographies or niches.
• **Redesigning business processes and organization structure** to reduce costs and increase flexibility.
• **Increasing productivity improvements** from technology investments. Lower cost of communications, computing, and the growing client/server environment has made decentralization far more feasible.
• **Using nonfinancial performance measures**, such as the number of customer complaints, number of on-time deliveries, relative share of defect-free units — all of which are critical to quality control and internal and external customer service.

Table 5–1 Cost Management Strategies

organization and each individual in the organization on quality and costs. The goal is managing costs of the process and quality/price of the output to meet the expectations of the customer. (See Fig. 5–1).

After identification and analysis of cost drivers, it is decided if there is a customer driven need for the output of the activity. Cost drivers that do not add value are isolated and eliminated. The continuous improvement efforts provide an ongoing control of quality and costs. Steps involved in process value analysis are identified in Table 5–2.

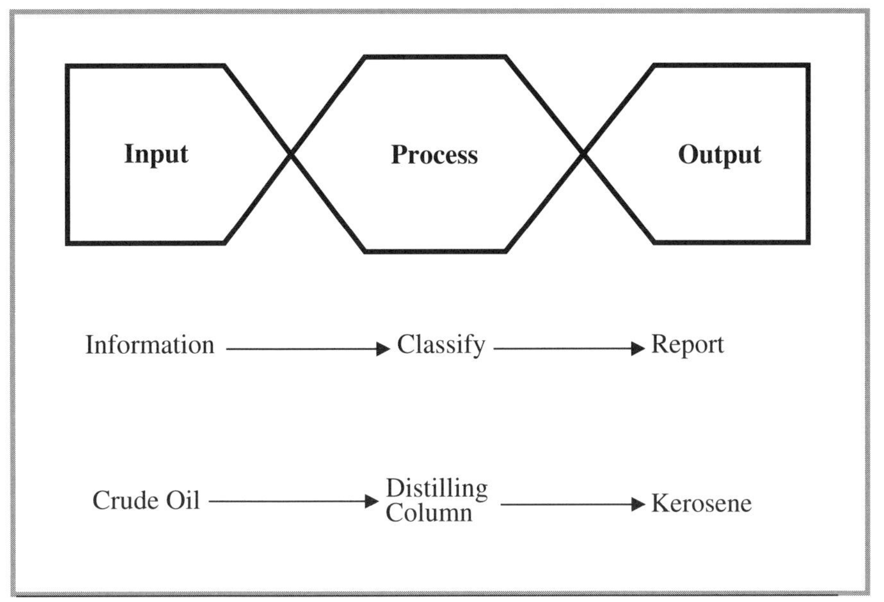

Fig. 5–1 Example of a Process

COST BEHAVIOR

Fundamental to the entire process of cost cutting is a clear understanding of how costs, particularly overhead or other expenses, should vary as related to business volume. A key to understanding cost behavior is distinguishing how costs behave with respect to changes in a particular cost driver. *Variable costs* vary directly and proportionately with the level of activity and in direct proportion to changes in a cost driver. *Semi-variable* costs have both fixed and variable behaviors.

An example of semi-variable costs is electrical energy expense. Electrical energy expense contains an element of both fixed and variable cost. The *fixed* or *demand* cost is established by a consumer's projected maximum demand for electrical energy. Investment by the supplier will be required to build their capacity to accommodate the user's maximum demand. The variable cost portion of the energy cost is based on the number of kilowatts used by the firm.

Fixed costs do not vary with the level of activity. They are the cost of resources invested in the activity for a given time. However, in the long-term fixed costs can be eliminated by divesting idle capacity or eliminating other drivers of uncontrollable costs. For several years, refineries in the oil industry had overcapacity. Refineries require large amounts of spending for new processes to meet environment requirements and customer's expectations. If the investment is not fully utilized profits are reduced by the cost of excess capacity.

Define the Process
- Document the process flow
- Define the input requirements for each process step
- Define the output of each step
- Identify customer (both internal and external) requirements
- Compare customer requirements with the input/output requirements
- Identify or define the required (full-time equivalent) staff level for each process

Analyze the Activities
- Analyze each activity within each process
- Using customer requirements, identify each activity as value-adding or nonvalue-adding
- Determine the cycle time for each activity
- Calculate for each process the cycle efficiency—the value-adding time as related to total time
- Cumulate the efficiency through the entire business chain

Analyze Cost Drivers
- Identify the cost driver—the cause and the effect
- Analyze the effect of eliminating the nonvalue adding drivers

Plan Improvements
- Determine by activity the costs of both the value-adding and nonvalue-adding activities
- Select methods to eliminate any nonvalue factors and optimize the value-adding ones

Table 5–2 Steps in "Process Value Added" Analysis

For short-term control purposes, costs are segregated into those that are controllable or non-controllable. The costs are further classified regarding behavior in relationship to activity, whether fixed, variable, or semi-variable as shown in Figure 5–2. When demand changes, the excess assets become sunk costs. In the long run, however, all costs become controllable as excess capacity can be sold or disposed of in other ways.

Costs with alternatives include *opportunity, incremental,* and *sunk*. *Opportunity* costs are opportunities foregone for another project. As an example, if management decides to retain earnings, there is an opportunity cost involved. Stockholders could receive the earnings as dividends and invest the funds in other stocks, bonds, real estate, or any other asset. The firm's earnings on retained earnings invested in operations must be at least as much as stockholders themselves can earn in alternate investments.

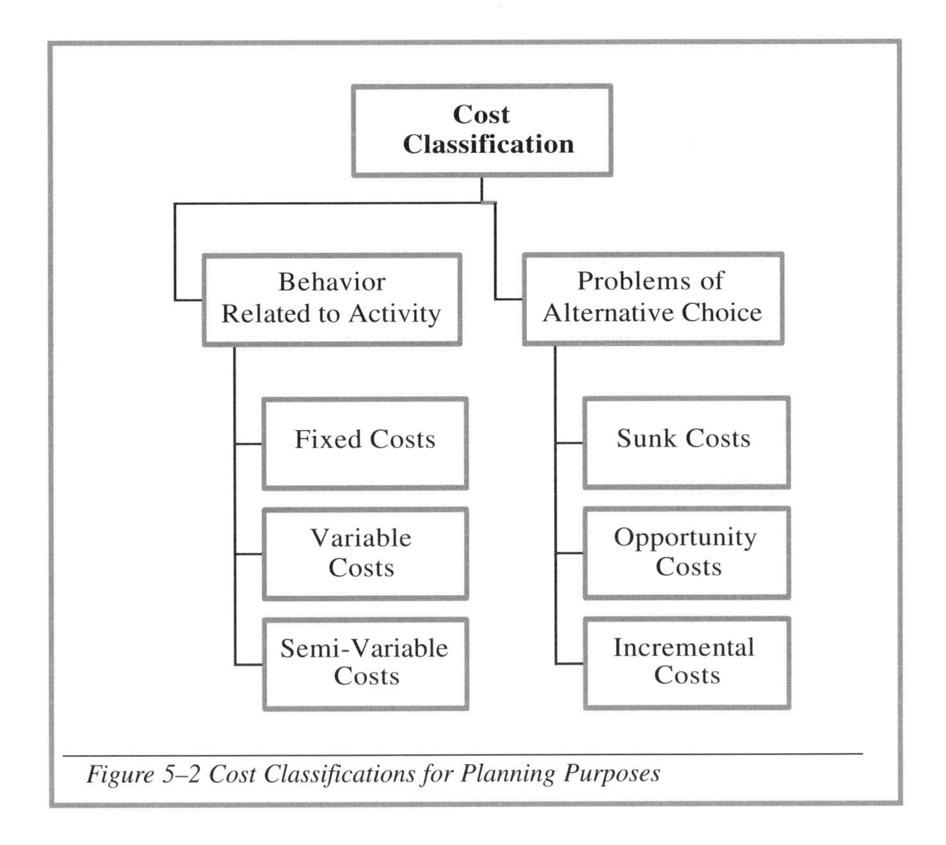

Figure 5–2 Cost Classifications for Planning Purposes

Incremental, sunk, and relevant costs are similar, yet have different impacts. The term, *incremental*, is widely used in finance. It refers to the change in total results under a new condition compared with some given or known conditions. Incremental cost analysis includes evaluating alternatives by using a "with and without" approach. That is, the general financial results "before the course of action," and "assuming the course of action is taken" results are compared to decide if net benefits result to the organization.

Sunk cost is really just another term for historical or past cost. As the sunk cost has already occurred, it becomes irrelevant to the decision making process. Alternatives for future opportunities should not consider sunk costs. An example of sunk cost is when demand decreases for a refinery. The overcapacity assets become sunk cost and should not be considered for future planning other than elimination of the costs.

For an example, IBC obtains a lease in Harris County, Texas. IBC drills a well 10,000 feet and finds oil. The costs of the well to date is $1.5 million and additional completion costs total an additional $.4 million. Projected production, oil prices, and operating costs are included in Fig. 5–3.

Year	Production (BBLs)	Price	Revenue	Costs	Operating Earnings before Taxes & DD&A
1	20,000	$20/bbl	$400,000	$100,000	$300,000
2	16,000	"	360,000	80,000	280,000
3	14,000	"	280,000	70,000	210,000
4	12,000	"	240,000	60,000	180,000
5	10,000	"	200,000	50,000	150,000
6	10,000	"	200,000	50,000	150,000
Total	82,000		$1,680,000	$410,000	$1,270,000

Figure 5–3 Decision Making Using Relevant Costs

The question now becomes should IBC complete the well since the completion costs will add $0.4 million to the $1.5 million for a total cost of $1.9 million for drilling and developing the well. If there is certainty that the well will produce 82,000 barrels, then the well should be completed as the total loss on the well will be reduced from $1.9 million by the $1.27 million income or a net loss of only $0.63 million.

If the additional completion work is not performed, loss on the well will be $1.5 million. Therefore the investment of $1.5 million is not relevant to the decision. Earnings above the $0.4 million incremental cost will be to the advantage of IBC. Sunk costs can sometimes be partially or totally recovered through write-offs, deductions, etc.

Financial managers frequently come under general criticism for a less-than-acceptable analysis of costs. If costs are planned and controlled properly, their true nature must be understood. What makes a cost increase or decrease with changes in activities or volume? What are the true drivers of cost?

Costs are managed in various ways throughout the oil industry but the key to success is recognizing that strategy should drive the design of key business processes, and that the company's organizational structure should facilitate—not impede processes. Companies are adapting corporate strategies to the realities of the new marketplace. Their primary goals are improved profitability and achievement of sustainable competitive advantage.

INTERNAL REPORTING OR CONTROL SYSTEMS

The management control system is a logical integration of management accounting tools to gather and report data and to evaluate performance. A well designed management control system aids and coordinates the process of making decisions and motivates individuals throughout the organization to act in concert. A management control system coordinates forecasting results, measuring and evaluating performance, and motivating employees. There are several aspects of a management control system. These include:

- Organizing the enterprise into subunits, or responsibility centers
- Establishing objectives for these subunits
- Quantifying and evaluating the performance of the subunits and communicating the performance of the subunits

Cost Center
The focus of the cost center is to control expenses and ensure they are within the limits set by the budgeting process. An example of a cost center would be a corporate finance department. Such a cost center will, for example, be concerned with the need to obtain the maximum productivity from employees, control employment costs, and provide quality service to its clients or customers.

Revenue Center
As the cost manager is responsible for the control of costs, so the revenue center manager is responsible for the production of revenue. A sales organization (national, regional, or local) is an example of a revenue responsibility center. While there may be costs incurred within the revenue center, control of these costs will be of minimal importance compared to the manager's responsibility for meeting revenue goals.

Profit Center
A step up in responsibility from the cost and revenue responsibility is represented by the profit center. The profit center manager is responsible for both revenue production and cost control within the center. An example might be a divisional manager who is responsible for the purchase of products (cost activity) from refineries and the subsequent sale of these products to consumers (revenue activity). Obviously, the responsibility of the profit center manager is greater than that of either the cost center manager or the revenue center manager.

Investment Center
The highest level of responsibility is represented by the investment center manager. The distinction between the profit center manager and the investment center manager is that the latter controls (and is responsible for) not only the profitability of the center, but also the amount of corporate resources invested in the activities of the center and the return on the resources.

Table 5–3 Definitions of Cost Centers

Each responsibility center has its own financial responsibility and management issues. The establishment of responsibility centers accommodates delegation allowing the decision-making process to take place at a lower level. Top management can devote attention to more global/strategic issues. The degree to which responsibility center managers are allowed to make decisions will vary between organizations. The concept, however, is the same.

There are differing types of responsibility centers and one or more will be found within an organization. The ordering of the list in Table 5–3 generally represents an increasing level of responsibility.

In a strategically managed organization, the objectives and goals of each cost center or group will support the goals and strategies of the total organization driven by the overall mission and purpose. The "balanced scorecard" performance measurement concept is a way to transform strategies so they can be designed more precisely and communicated into the organization as objectives. Direct communication can translate strategy into a set of action-oriented performance measurements.

The Relationship between the Management Control Process and Transfer Pricing

The management control process (MCP) has two major objectives: (1) guiding members of the company toward company goals, and (2) evaluating the progress of the company's segments toward these goals. These goals are further defined in Table 5–4. Transfer pricing is a system within the MCP. Viewed in this manner, transfer pricing involves more than simply determining the amount of the transfer price.

After developing the major plan for the organization, factors that influence transfer pricing decisions are determined. The main factors are managerial, such as desire to use profit centers rather than cost centers, tax considerations, influences of a host country's government, or the type of market faced by a company. A consideration of these and other factors often leads to a decision to use certain transfer pricing techniques in the organization. Transfer pricing techniques are then communicated to the performance centers for use in calculating actual transfer prices.

The financial role in the transfer pricing system involves:

1. Providing input for the selection of the transfer pricing techniques
2. Monitoring transfer pricing throughout the company, and
3. Measuring the results of transfer pricing via reports on the performance of company segments

These decisions on transfer pricing cannot be made without consideration of the major objectives of the MCP.

Objective 1
To guide the members of the company toward the organization's goals (Goal Congruence). Suppose that one goal of a production company is to minimize the risk of environmental pollution. How does the organization encourage managers to accomplish this goal? Survey of existing facilities for potential problems can be required. Targets can be set for investments in pollution control equipment. Mandatory educational systems can be established so that managers are aware of potential pollution problems, alternative solutions, and government regulation. In other words, standards are established.

Objective 2
To evaluate the progress of the company members toward the goals. Management needs to determine if the goal of minimizing the risk of environmental pollution is being achieved. Information must be collected. Has the survey of existing facilities been completed? What are the results? Was the survey done well? Are the targets for investments being met? Are managers attending the educational classes? Are the classes successful? In other words, actual performance is compared to the established standards, and, if necessary, corrective action is taken.

Table 5–4 Objectives of the Management Control Process. (MCP)

The transfer pricing technique selected by a company must simplify the objectives of goal congruence and performance evaluation. To a large extent, the form of the MCP is dictated by the structure of the organization. Structure as used here does not refer to the organizational chart. It refers to the extent of the segment manager's responsibilities and degree of interdependence among the segments. The extent of responsibilities is related to decentralization. The degree of interdependence is related to differentiation. The extent of decentralization and differentiation determines the amount of integration, or segment coordination, within an organization.

The following illustration demonstrates goal congruence and performance evaluation:

When gasoline is produced by ABC Oil Company's refinery, coke is created as a byproduct. Because the volume of coke sales will be small compared to other major products and because coke has a narrow profit margin, the marketing unit normally would not choose to expend any effort selling the product.

To encourage the marketing unit to sell the product, ABC Oil Company establishes a price for the transfer of coke from the refinery to marketing. A low price will give marketing a large profit margin and make sales attractive. The refineries can, therefore, be assured of an outlet for a product they create in producing gasoline. In addition, corporate resources are not invested in large inventories of unsold coke. The transfer price encourages the business units to work together toward a common corporate goal.

Few problems arise in decentralized organizations when all the segments are independent of one another. Segment managers can focus on their own segments without hurting the organization as a whole. In contrast, when segments interact greatly, there is an increased possibility that what is best for one segment hurts another segment badly enough to have a negative effect on the entire organization.

There is a significant interdependency within an integrated oil and gas company. The production company transfers crude to the refineries where the crude is processed into products and then transferred to the marketing divisions. In the natural gas area, the production company transfers the gas to a gas plant or transmission company, which in turn transfers gas to the marketing divisions or local distribution companies (LDC). Transfer prices guide managers to make the best possible decisions regarding whether to buy or sell products and services inside or outside the organization. Another important reason for evaluating segment performance is to motivate both the selling manager and buying manager toward goal-congruent decisions.

In a survey of transfer pricing in large integrated companies, 85% of the responding firms reported that they used transfer pricing. In the responding firms, the transfer price was determined by:

57%	*Cost*
30%	*Market*
7%	*Negotiated*
6%	*Variable Cost*

The rationale for using transfer pricing included:

47%	*Profit*
21%	*Cost determination*
23%	*Control and accountability*
9%	*Other*

Cost-Based Transfer Prices

Approximately half of the major companies transfer items at cost. However, there are numerous definitions of cost. Some companies use only variable cost, others use full cost, and still others use full cost plus a profit markup. When the transfer price is some version of cost, transfer pricing is nearly identical to cost allocation. Costs are accumulated in one segment and then assigned to another segment.

Transferring or allocating costs can disguise a cost's behavior pattern. Other problems arise if actual cost is used as a transfer price because actual cost cannot be known in advance and the buying segment lacks a reliable basis for planning. Inefficiencies are merely passed along to the buying division and the supplying division lacks incentive to control its costs.

In some companies, a version of activity-based costing (ABC) is used to prevent disguising a cost's behavior pattern. In ABC, drivers of cost are identified and agreed upon by both parties. Only those costs generated by agreed upon cost drivers are allocated to the recipient.

Market-Based Transfer Prices

If there is a competitive market for the product or service being transferred internally, using the market price as a transfer price can lead to the desired goal congruence and managerial commitment. The market price may come from published price lists for similar products.

For transfers within the oil and gas industry, there are posted prices for the products transferred. Posted crude prices plus published transportation tariffs are available for pricing crude transfers to the refineries. Rack prices plus published transportation tariffs are available for pricing transfers of products from the refineries to the marketing companies. And, of course, competition and market prices govern the sales price to the consumer.

Before the unbundling of the natural gas industry, production companies sold to transmission companies based on take-or-pay contracts. The transmission companies in turn sold to the marketing distributor based on a transportation tariff plus cost of gas. FERC's (Federal Energy Regulatory Commission) Order 636 has completed the unbundling of transmission companies and now producers can ship to the consumer or marketing arm of the gas industry.

Variable Cost Pricing

Sometimes market prices do not exist. In these instances, versions of cost-plus-profit are often used in an attempt to provide a fair or equitable substitute for regular market prices.

Services are frequently provided to affiliated units at cost plus overhead rate. The overhead rate covers indirect costs plus a profit markup. In addition, because of scarcity or incentive prices, materials may be centrally purchased by one segment and provided to other segments. There was a scarcity of pipe in the early 1990s and, because of environmental requirements, pipeline companies sometimes placed significant orders for a particular mill run and supplied the other segments of the company. This provided a special price and ensured adequate availability of pipe.

Negotiated Transfer Prices

Companies heavily committed to segment autonomy often allow managers to negotiate transfer prices. The managers may consider both costs and market prices in the negotiations, but no policy requires them to do so. Virtually any type of transfer pricing policy can lead to dysfunctional behavior—actions taken in conflict with organizational goals.

Top managers who wish to encourage decentralization will often make sure that both producing and purchasing division managers understand all the facts, and then allow the managers to negotiate a transfer price. Well-trained and informed segment managers who understand opportunity costs and fixed and variable costs will often make better decisions than will top managers.

INTERNATIONAL TRANSFER PRICING

Development of the Transfer Price

It is usually the responsibility of the corporate controller to develop the transfer prices that will be used in dealings between subsidiaries or between parent and subsidiaries. The price used will result in a profit or loss for the exporting company and will directly influence the profit or loss ultimately realized in the importing company.

Such a situation has led some multinationals to develop transfer prices that are very subjective and result in the lowest profit (or even losses) being reported in the highest tax rate countries. Conversely, the highest profit is reported in the lowest tax rate countries.

While this approach seems logical and defensible at first glance, it is unfair. The result is more laws and restrictions in the countries that are being taken advantage of. This in turn has adversely affected the multinationals in terms of compliance expenses and the reduction of free trade. The use of subjectively developed transfer prices also complicates the measurement system. This frustrates local management because it cannot control or influence a significant item of cost.

A more businesslike approach is for the controller to establish a uniform and objective system of transfer pricing that is:

- Simple and inexpensive to administer
- Understandable to the users, and
- Fair in its effect on measurements

An arm's-length, objective system can take two forms:

- *Cost-based system* favored by American companies uses the output of the company's cost accounting systems and adds a markup for profit. The markup on the transfer of services performed by one unit for another is more subjective. Flat percentages are usually established and not changed.
- *Market-based system* favored by many European countries starts with the expected selling price in the importing countries. This is then reduced by estimated locally incurred cost, expenses, and profit to arrive at a transfer price that is affordable to the importing company.

Royalties and Other Fees

Whether using a cost-based price system or a market-based transfer price system, the multinational company will probably also employ a "royalty or other fees" system. The transfer price system is used to move goods and services between related companies. A *royalty system* is used to recover, from the subsidiaries, compensation for know-how, the use of brand names, and company logos, etc. The two most common royalties charged to subsidiaries are:

- Research and development
- Parent-company management

The royalty for research and development is traditionally a percentage of sales. Usually a parent-company management royalty is not based on a ratio. Rather, it is an apportionment of the total parent-company staff and general expense based on a relationship of the subsidiaries. One method is to apply the percentage each subsidiary contributes to gross income to the total parent-headquarters expense pool. Another is to negotiate a fee based on number of employees, profit margin, growth, etc., that recovers all parent headquarter's expenses.

The advent of U.S.-based multinational corporations and their continued growth have added complicated dimensions to transfer pricing. The complications arise because international transfer pricing must meet not only the objectives of the management control process, but also a wide variety of other objectives. These other objectives are sometimes so important that the objectives of goal congruence (profit maximization) and performance evaluation are considered secondary or unachievable.

Income Tax Implications Affecting Transfer Prices

By far the most persuasive objective in international transfer pricing is tax minimization. The economic advantages are obvious and immediate if a transfer price shifts profit from a country with a high tax rate to a country with a low tax rate. For example, if a product is being manufactured in a country with a favorable tax structure, a high transfer price will result in more of the profit from the sale of the product being taxed at the favorable rate.

The other objective, performance evaluation, is even more difficult to attain when the objective of tax minimization is pursued. The profit in one country will be greater than the profit in another country, not because of better management but because of the transfer price. One way around this problem is to establish a separate reporting system for control purposes so performance evaluation is based on more realistic reports.

To calculate how much tax a producing company should pay, a government needs to know the value of the oil sold. In the early years, a company might sometimes hoodwink a government by underpricing its oil. Then it would sell it to a wholly owned trading company domiciled in some low-tax or zero-tax country. The trading company would then inflate the price and sell the oil to its refineries in consuming countries. Through this process, profits were minimized in the taxable producing and refining businesses, but maximized in the tax-free countries. For obvious reasons, this has been labeled transfer pricing abuse, and it is now seldom attempted. Even so, governments have not forgotten the days when such abuse was commonplace, and they remain quite suspicious of the oil prices used in oil companies' tax returns. To resolve this issue, different governments use different techniques:

1. *Declaration.* The oil company has to declare to the government the price at which the oil was sold, and swear that it was an arms-length transaction (i.e., not a contrived deal to minimize taxes). The government monitors all the declarations it receives, and it investigates any that seem to be out of line. Making a false declaration can be a criminal offense with a severe penalty.
2. *Tax reference pricing.* The government deems a price and taxes companies on that basis. If the deemed price is too high, and the producing company is unable to sell its oil at anything like the deemed price, then "government take" can easily exceed 100%. If a company in this position is unable to convince the government of this, its ultimate option may be to shut-in production.
3. *Formula pricing.* The oil company and government agree on a basket of crudes that approximates the value of the production stream to be taxed. For example, both parties may agree to value a particular crude according to the formula: 25% Dubai + 25% Brent + 25% Alaska North Slope + 25% West Texas Intermediate Sweet Crude – $1.20. This might be calculated and applied one month retrospectively. The important thing is that the crudes in the basket should be widely and transparently traded in large volumes.
4. *Auction.* A government may auction its own crude entitlement and use the price obtained as a basis for taxing the oil companies.

Which technique a government chooses to use will depend on many factors. Use of the declaration method is likely to be unwise if the government has any doubts about the incorruptibility of its middle-level civil servants. But whichever method is used, a company operating, or planning to operate, in the country must pay careful attention to this matter. An unfair method of tax pricing can convert an apparently benign government regime into an excessive one.

Netback Pricing

There is an assumption that the producer drives the market. This is not always the case. Sometimes a producer will make an agreement to a refinery marketer to sell their products and the producer will guarantee a profit margin. The formula is:

Crude Price = Income from sale of products
– Refining cost
– Transportation cost
– Refiner/marketer's profit margin

Import Duties

Import duties, as well as income taxes, can be minimized. For example, a company benefits economically if it transfers products at low prices to a country with high import duties. All things being equal, the company can reduce the total cost for import duties through transfer pricing. Although import duty minimization sounds easy, frequently it is complicated because "all things" are rarely equal. For example, a country with low import duties may have high income taxes, or a country with high import duties may have low income taxes. Another factor is the tax rate of the country from which the product is shipped. Thus, the company must consider import duties along with income taxes in both the shipping and receiving countries.

Minimization of income taxes and import duties is an important goal. Recently, however, taxing authorities in many countries have begun to pay closer attention to attempts by companies to transfer profit to countries with lower taxes. Overzealous efforts to minimize import duties and taxes may result in short-run gains but long-term losses.

A second economic restriction by some countries is disallowing certain expenses against taxable income. For example, some general administrative expenses or research and development expenses may be disallowed if they are performed elsewhere. Another example is royalty fees charged by management against subsidiary income. To the extent these are disallowed by the host country, the amounts can be recaptured by increasing the price of goods shipped into the country.

Currency Fluctuations

During periods of currency instability, the performance reports of foreign affiliates can be affected dramatically by exchange rate fluctuations. Many United States-based multinational companies find it convenient to evaluate the performance of foreign affiliates with reports stated in U.S. dollars. If currency exchange rates fluctuate during the performance period, however, it may be difficult to evaluate the performance of the affiliate. At the same time, management of the affiliate often finds it more convenient to evaluate its performance with reports stated in local currency rather than U.S. dollars.

JOINT OPERATING INTERESTS

The transfer of costs is somewhat different in joint operating interests, production sharing agreements, concession or other sharing agreements. The transfer procedures are usually covered by the operating or concession agreement. Within the operating agreement are various exhibits; one is the accounting procedure—the primary instrument used by accountants to determine the proper charges to the joint account.

Charges to the joint interest by the operator are usually classified as direct charges, overhead, pricing of joint account material purchases and inventories. Overhead or indirect expenses are charged to the interest by use of an allocation of district expense plus a fixed rate for administrative overhead, plus warehousing charges. Generally, accumulated charges in the district expense account are apportioned to all properties served on some equitable basis consistent with the operator's accounting practice. This gives the operator some flexibility; however, this procedure must be equitably and consistently applied.

When drilling both onshore and offshore wells in the same district, some basis of cost allocation is required. Naturally, drilling wells should attract a greater allocation than producing wells, and conversely, offshore drilling and producing wells should attract a greater allocation than onshore drilling and producing wells.

Parent Company Operations (PCO)

An operation with its affiliate office in Cairo might have its parent company office in Chicago or Houston. Every affiliate has a parent company office. It is universally agreed that the parent company offices cost money and that the operator should be permitted to collect a contribution from its partners.

The problems occur when trying to agree on what is and what is not covered by the PCO fee. At one extreme, it is easy for all to agree that the fee covers monitoring by senior management and that the operator will not attempt to make any additional charge for that. At the other extreme, it can usually be agreed that the operator will be allowed to make additional charges to the joint venture for technical work done by parent company specialists, against formal documents known either as authorization for expenditure (AFEs) or as service work orders (SWOs). Items covered by SWOs might typically include special reservoir studies, legal work to counter a particular suit against the joint venture, and so forth.

Middle ground topics between these clear extremes cause the problems. For example:

- *Manpower capacity charges.* A major will typically keep a reservoir of skilled and experienced people in its head office, ready to move quickly as needs and opportunities become apparent. Such talent warehouses are one of the great but costly strengths possessed by major companies. A major company operator will argue that its partners benefit from this but do not pay for it. To remedy this, the operator may try to levy and annual fee of say $20,000 per person sent to the joint venture from the head office. With 100 expatriates, this fee could substantially exceed the normal parent company overhead.
- *Technology recovery charges.* Many of the majors have large and sophisticated laboratories working on exploration and production (E&P) topics. A major company operator is likely to argue that its partners benefit from the availability of this technology and ought to pay some kind of standby fee for it, as well as the direct cost of experts and material included in any approved SWO.

Overhead Allocation

An operator may be responsible for running a number of different joint ventures within the affiliate, each with its own unique group of partners. Part of the cost of the affiliate head office is likely to be for the affiliate's own account. But much of the office cost will have to be allocated among various ventures in an equitable way. There are various alternatives:

- *Time sheets.* Professional-level employees in departments that work directly on joint-venture activities are required to complete time sheets. Such departments include exploration, drilling, engineering, accounting, purchasing, and so forth. Employees above and below this professional level and employees in staff departments that have only indirect impact on any particular joint venture are not required to keep time sheets. The cost associated with each time sheet hour is then multiplied by a factor that exactly liquidates all non-timesheet costs. This gross hour figure is then charged to the relevant joint venture to the operator's own account. For many operations, the multiplier will be in the 3.0 to 4.5 range.
- *Capital Investment.* In this method, the amount of capital investment being made in any joint venture is accepted as a measure of how much affiliate-office effort is being devoted to that joint venture. The total office cost is, therefore, allocated among the joint ventures in proportion to their capital expenditures. This has two important advantages for the operator. First, the method is very simple to apply; second, the operator's own 100% activities are unlikely to involve much capital investment and will therefore attract little if any of the overhead.

- *Other Methods.* These include allocation in proportion to total expenditure or production (i.e., a per barrel basis).

Use of Independent Experts

Many Joint Operating Agreements contain a provision to use an independent expert to resolve differences of technical opinion that may arise among the partners. The challenges of the nonoperating partner are to exert pressure on the operator for improved performance. The key question is how much time, effort and money are necessary to invest in this activity. A considerable amount of academic work is now being done on this problem, under the loose title of "agency theory". At present, agency theory is too rudimentary to help the nonoperator with its problem, but this may soon improve and it is an academic field worthy of close attention.

There are likely to be four main determinants:

- The level of influence that the nonoperator can exert
- The trustworthiness of the operator
- Cost incurred by the nonoperator when monitoring the operator's activities, and
- Materiality

The type of work nonoperators perform:

- During development, recommend different and better kinds of development alternatives
- Provide useful comparisons of other joint venture performance.

ASSIGNMENT OF COSTS TO RESPONSIBILITY CENTERS

One of the critical issues in the use of responsibility centers as a control tool is the assignment (allocation) of costs to the subunits of the organization. If costs are not assigned in an effective fashion to a responsibility center, then a cost center manager may be held responsible for costs that he or she cannot control. Not only does this mean that cost control will be ineffective, it also suggests that the manager will view the budget assigned to the center as unrealistic and (probably) unachievable.

For example, a production line supervisor may have direct labor hours as an item for which he is accountable. It would be a mistake, however, to include in the supervisor's cost center items over which he has no control. Such items would include the cost of utilities to air condition the building or the administrative costs of the plant. Such costs should be included in the responsibility center of a higher level manager (e.g., the plant general manager) since this manager has more control over such costs.

There are two issues that arise in addressing this matter. One is determining those costs that are truly controllable by the individual manager. The other is deciding whether certain indirect costs (i.e., those not directly associated with a particular responsibility center) should be allocated to units within the organization. We can think of these concerns as related to the issues of:

- Cost controllability
- Cost allocation
- Activity-based costing
- Contribution margin

Cost Controllability

Management control systems often distinguish between controllable and uncontrollable events and costs. Determining whether a specific cost/expense is controllable by a responsibility center manager can be a difficult task. Insight can be gained by asking the following questions:

- Does the manager have the authority to cause the cost to increase or decrease?
- Is the cost specifically associated with a particular unit's activities?

It may be helpful to consider the issue of controllability in light of direct and indirect costs. A *direct cost* is one that is associated with a specific activity. One observation is that all controllable costs are direct costs, since a direct cost will by definition be associated with a particular cost driver.

While controllable costs are a subset of direct costs, not all direct costs are controllable by the responsibility center manager. For example, the lease worker's labor rate per hour is a direct cost with respect to a production supervisor. The supervisor, however, does not control this cost. It may, for example, be negotiated by the management of the company with a labor union.

Furthermore, not all indirect costs are controllable. Since *indirect costs* are by definition not related to a specific activity, they can not be traced to a specific responsibility center.

Cost Allocation

A related issue is whether indirect costs should be allocated to responsibility centers. Indirect costs, by definition, are not controllable by responsibility managers. At first glance, it may seem that it is therefore inappropriate to allocate the costs to the business units.

However, in many cases, responsibility centers benefit from the service provided by an indirect cost. For example, the expense of a corporate tax department is not controllable by a division manager. Nonetheless, the division benefits from the activities of this indirect cost. In the absence of such a group, it would be necessary for the division manager to either hire someone to perform all tax functions for his business unit or to outsource the work. Since every division manager would be faced with the same choice, the resulting approach would be less organized and less cost efficient.

While it is generally agreed that allocation of indirect costs is not effective for decision-making purposes, it is controversial whether allocation of such costs should be undertaken for control purposes.

Perhaps the most important reason for not allocating such costs is that the indirect costs are not controllable by the responsibility center manager. Consequently, such an allocation tends to undermine the effectiveness of the responsibility center as a management control device and makes the unit manager feel that he is being held responsible for costs he cannot directly control.

There are reasons for allocating indirect costs to responsibility centers. First, it demonstrates the true requirements for profitability of the organization. As illustrated in the above example, the need to comply with the requirement to file tax returns and otherwise deal with government regulations is simply a cost of doing business. Failure to recognize this cost presents a false picture of the profitability of individual products and ultimately of the entire organization. A second reason is that such allocations cause managers to use allocated services judiciously (to the extent that use of the service is under the discretion of the manager).

For example, copying services, computer services and telecommunication services are typically charged to responsibility centers on the basis of usage, at least to some extent. (There may be a fixed charge allocated to all units, with a variable charge based on actual use). If the responsibility center manager is responsible for this allocation, then he will do everything he can to control this cost. Arguably, this results in a more cost effective use of the corporate resource.

Production companies often are required to allocate indirect operating costs to wells and leases because of the need to calculate netback at the well for either the royalty owners, production sharing agreements, joint ventures, tax calculation, severance taxes, etc. The basis for allocations of indirect costs are generally outlined in operating agreements, tax laws or industry practice. The Council of Petroleum Accountants Societies (COPAS) in the U.S. has developed extensive data for guidelines in the allocation of indirect costs.

Method of Allocation

If the decision is made to allocate indirect costs to responsibility centers for control purposes, the next question is how this should be done. First, to the extent possible, allocation should be based on the usage of the service represented by the cost. The telephone expense or copying expense discussed earlier is an example.

Other allocations may be based on the relative benefit provided to the center. For example, cost of the personnel function may be allocated to a unit based on the number of employees in the unit, as a percentage of the total number of employees in the corporation. Rent expense on a building may be allocated based on the relative amount of floor space occupied by a particular responsibility center. Other costs (such as top management salaries) may be allocated to units on the basis of relative revenues, because it is difficult to determine how much each unit benefits from this expense.

It is generally desirable to allocate indirect costs to the responsibility centers based on the budget for the indirect services, rather than the actual cost of the services. Otherwise, inefficiencies in the operation of the indirect activity will effectively be passed on to the operating units. In practice, it may be difficult to allocate only the budgeted amount, since the excess of actual indirect cost over the budgeted amount may be due to increased activity on the part of the operating units.

	--$MM-- Segments			
	A	B	C	Total
Revenues	$200	$150	$300	$650
Less variable and direct operating expenses	100	75	150	325
Equals Contribution Margin	**100**	**75**	**150**	**325**
Less fixed costs controllable by segment manager	75	50	100	225
Equals Contribution controllable by segment manager	**25**	**25**	**50**	**100**
Less fixed costs controllable by others	15	15	40	70
Equals Contribution by segment	**10**	**10**	**10**	**30**
Minus unallocated costs				6
Equals Company's Income before Income taxes				**$24**

Table 5–5 Contribution Margin Reporting

Contribution Margin

Many organizations combine the contribution approach to measuring income with responsibility accounting. That is, they report by cost behavior as well as by degrees of controllability. Contribution margin is revenues less variable or direct operating expense. The format of the income statement would take the format illustrated in Table 5–5.

Segments are responsibility centers for which a separate measure of revenues and costs are obtained. In the above format each allocated line of the income statement is reported by segment.

HOW A TRANSFER PRICE CAN AFFECT PROFIT MAXIMIZATION

If a refining division sells gasoline to an independent marketer, the division will incur $300,000 in allocated marketing costs. If the refining division transfers the gasoline to the marketing division, the marketing division will spend $2.4 million on processing and marketing activities and then receive $5.8 million for the 200,000 barrels at the service stations. A selling price to independents has been set at $18 per barrel, while the transfer price to the marketing division has been set at $16 per barrel (Fig. 5–4).

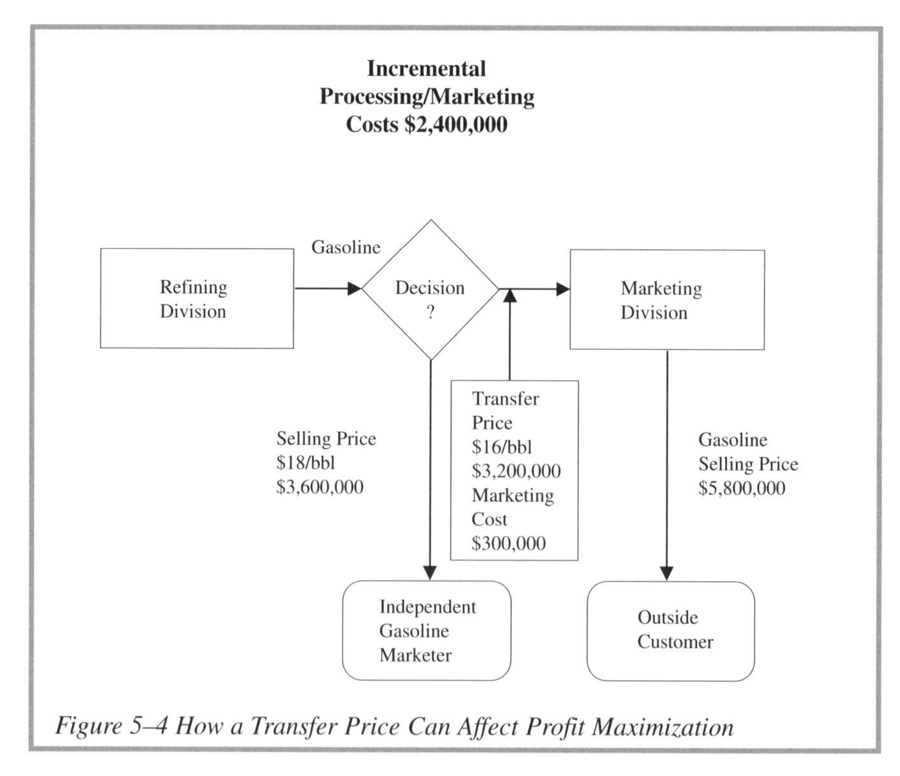

Figure 5–4 How a Transfer Price Can Affect Profit Maximization

The refining division manager would base his decisions on the following relevant information:

	Sale to Independent Gasoline Marketer	*Transfer to Marketing*
Revenues	*$3,600,000*	*$3,200,000*
Marketing Costs	*300,000*	*-0-*
Profit Contribution	*$3,300,000*	*$3,200,000*

Because his division can earn $100,000 more by selling the 200,000 barrels to an independent gasoline marketer, and assuming that the division is evaluated by its level of profits, the refining division manager would decide to sell the gasoline to the independent marketer.

An individual who had responsibility for both the refining and the marketing divisions and who used all the information would reach a different conclusion. If the interests of the entire corporation are considered, the alternatives appear different:

	Sale to Independent Gasoline Marketer	*Sale to Ultimate Consumer through Marketing Division*
Revenues	*$3,600,000*	*$5,800,000*
Costs:		
Marketing by Refining Division	*– 300,000*	*-0-*
Processing & Marketing by Marketing Division	*-0-*	*– 2,400,00*
Profit Contribution	*$3,300,000*	*$3,400,000*

The corporation as a whole can earn $100,000 more by absorbing the marketing division's $2.4 million in marketing and processing costs, not incurring $300,000 in marketing costs in the refining division and selling the product directly to the consumer through company-owned gasoline stations

Performance Evaluation in Profit Centers

The performance evaluation of a profit center depends on the difference between revenues and expenses. Because transfers out of a profit center are part of the revenues of the supplying segment and transfers into a profit center are part of the expenses of the receiving segment, the transfer pricing technique selected will affect the profits of both the supplying and receiving segments. The prudent segment manager will seek to transfer in products at as low a cost as possible and transfer out products at as high a price as possible. To the extent this is successful, the profit of his or her segment will be improved.

MOTIVATION, PERFORMANCE AND REWARDS

The management control system must recognize and accommodate the importance of the behavioral models. Using the motivational criteria of congruence and effort, top management chooses responsibility centers, performance measures and rewards. Incentives are defined as those informal and formal performance-based rewards that enhance managerial effort toward organization goals. Numerous performance measures have been described in previous chapters—whether to measure divisional performance by contribution margins or operating incomes and whether to use both financial and nonfinancial measures of performance. Managers tend to focus their efforts in areas where performance is measured and where their performance affects awards. Thus, accounting measures which provide relatively objective evaluations of performance are important.

Rewards may be both monetary and nonmonetary. Linking rewards to performance is desirable but often a manager's performance cannot be measured directly. Managerial performance and responsibility center results are certainly related, but factors beyond a manager's control also affect results. The greater the influence of noncontrollable factors on responsibility center results, the more problems there are in using the results to represent a manager's performance.

SUMMARY

Numerous challenges face the oil and gas industry in the realm of cost management. Cost management begins with the organization and the attitudes of members of the organization. The process-focused organization is inclined to identify and meet the needs of the customer rather than focus upon individual specialization which is costly to an organization. The process-focused organization is inclined to be more flexible and can meet the needs of changes and cost reductions.

Decentralized versus centralized organizations require certain cost management procedures. The continued thrust toward globalization requires special cost allocation and revenue procedures to assure host governments that the company is indeed a world player and will play fair. The traditional shared interests focused on accommodating all of the players in the venture fairly. This has trained the oil and gas industry to be aware of the legal requirements of different types of players whether it be a nation, a company, or an individual.

"Motivation"

The obvious objective of the company is to meet the needs of its customers, partners, owners, and employees. Rewarding employees for accomplishments reinforces behavior that will continue to work for the good of all shareholders.

Capital Budgeting and Investment Theory

6

The basic theories of economics and risk analysis are part of the fabric of our everyday lives and the concepts are virtually timeless. However, with the computer age and the age of capital, the sciences of economics and decision theory have evolved to a lofty position in the business world. The language of risk analysis can be rather exotic and confusing at times. Understanding the language and terminology of these theories allows professionals to communicate, yet sometimes a number of terms are used for the same concept.

Even some of the more complicated sounding theories are really concepts that are familiar to nearly everyone. Once the terminology is understood, an understanding of the theory is relatively easy. Nearly every facet of economic or financial analysis deals with one or more aspects of value. Value is the measure by which companies measure corporate wealth and whether or not the organization is growing.

Investment Theory

The foundation stones of investment theory are Present Value Theory and Expected Value Theory. They go hand in hand. They are used to estimate and characterize the value of assets, liabilities, stock, bonds, and investment opportunities. Present value theory recognizes the importance of the time value of money. The objective of the present value concept is to discount future cash flows to current value and compare with up-front costs required to be invested in the project. All other things being equal, the project with the most excess discounted cash flow over invested costs will be preferred over projects with less excess cash flow.

Fundamentals of Valuation

The concept of value can be viewed many different ways. The perspective is different for bankers, accountants, shareholders, management, regulatory agencies, and for buyers and sellers. Most engineers, analysts and shareholders focus on two general concepts of value: *market value* and *fair market* value. Understanding the true value associated with any investment option is fundamental to the corporate decision-making process.

Fair Market Value

Fair Market Value (FMV) is the price at which an asset would pass from a willing seller to a willing buyer after exposure to the market for a reasonable period of time. It is further assumed that both buyer and seller are competent and have a reasonable knowledge of the relevant facts, and that neither party is under any undue compunction to buy or sell.

FMV is not a valuation technique. It is a concept, based on value that can be derived by several techniques. The relationship between value and price can be quite complex. The difference between price and value often reflects an increase (or decrease) in corporate wealth.

Another relationship between value and price is called the *winner's curse*. It is important to keep this relationship in mind especially in competitive bidding situations. Figure 6–1 illustrates this concept. Theoretically, the average estimate or bid would most closely approximate the true value of an asset. However, it is not the average bid, but the highest bid that succeeds in a competitive bidding situation. Some define winner's curse as the difference between highest bid and the next highest bid. Others define it as the difference between highest bid and the average bid.

Profitability Criteria

There are a number of criteria used in the industry to quantify and characterize the value of an asset or a producing property. Some are superior to others and some are a waste of time. Sometimes management uses yardsticks that are virtually obsolete and misleading.

There is no single measure of profitability that can fully characterize the nature of the value of an asset. Numerous criteria exist and regardless of their relative value, an evaluator must understand the process and its strengths and weakness. Some common measures used to quantify and determine value of an asset are:

- Payout
- Capitalized Cash Flow/Cash Flow Multiple
- Accounting Rate of Return
- Profit-to-Investment Ratio (P/I)
    - Undiscounted (most common)
    - Discounted P/I Ratio (less common)
- Net Present Value (NPV)
- Internal Rate of Return (IRR)

Payout

Payout is the length of time (usually expressed in years) that it takes for net cash flow to equal the initial capital investment. The concept of return of capital or "cost recovery" is universal.

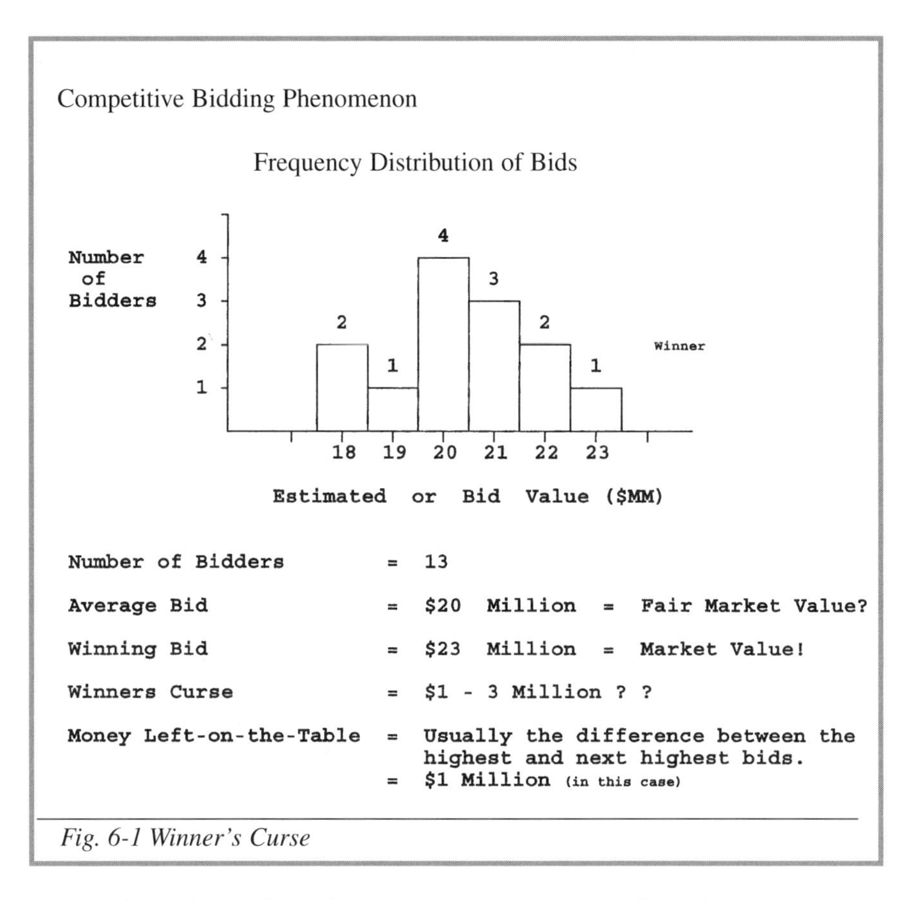

Fig. 6-1 Winner's Curse

From a financial point of view, the quicker the payout or recovery of costs the better. Payout typically refers to the time in which the firm gets its initial investment back and is a measure of risk.

The payout measure is used widely. In fact, it is often overused because it has some significant weaknesses and there are better measures. However, in management presentations if payout is not explicitly addressed then someone will raise their hand and ask, "When is payout?"

Cash Flow Multiple

When the information is available, a multiple of cash flow provides a rough estimate of the value of producing properties. The market for oil and gas production in the U.S. will typically pay from four to six times (annual) cash flow for a mature portfolio of oil or gas production.

This corresponds to payouts from five to seven years. Payout will typically be greater than the cash flow multiple for a production acquisition because production and cash flow are characterized by a declining profile.

"The good news is, the bad news is..."

The cash flow multiple is not quite the same as the often-used term—payout. *Payout* is the amount of time it takes for an investor to get his investment back. If a producing property did not have a decline rate, the cash flow multiple and the payout would be the same.

Profit-to-Investment Ratio

The Profit-to-Investment ratio is one of the oldest and most fundamental investment concepts. It will probably always be part of analysis and presentations. In a few cases the time value of money is not considered; however, generally the discounted future cash inflows will be compared to cash outflows. It is simply a comparison of the discounted cash flow that is ultimately received to the investment that earned it. The P/I ratio provides a ranking index to be used when numerous projects are considered in the capital budgeting process. Those with a higher index will take preference over discretionary projects with a lower index.

Profitability Criteria	Abbreviations/Synonyms Other Approaches	
Payout		Cost Recovery Period
Capitalized Cash Flow		Cash Flow Multiple
		Price/Cash Flow Ratio
Profit-to-Investment Ratio (P/I)	(P/I)	Profit/Investment Ratio
	(ROI)	Return on Investment
	(ROR)	Rate of Return
		P/I Ratio
		Discounted ROI
		Leverage
		Percentage Payout
	(PI)	Profitability Index
	(ROI)	Cash-on-Cash Return
Net Present Value (NPV)	(PV)	Present Value
	(PVP)	Present Value Profit
	(PW)	Present Worth
		Present Worth Profit
	(Discount Rate should be specified, such as NPV 15%)	
Internal Rate of Return (IRR)	(ROR)	Discounted Rate of Return
		DCF Rate of Return (DCFROR)
		Internal Yield
	(PI)	Profitability Index
		Marginal Efficiency of Capital
	(DCFR)	Discounted Cash Flow Return
Growth Rate of Return		Earning Power
	(MRR)	Modified Rate of Return
Accounting Rate of Return		The Baldwin Method
		Appreciation of Equity Rate of Return
		Simple Rate of Return
		Unadjusted Rate of Return

Table 6–1 Valuation Processes and Variations in Terminology

PRESENT VALUE THEORY

The most fundamental financial concept is the *time value of money*. The old adage is true, "A dollar today is worth more than a dollar tomorrow." The difference between the value of a dollar today and a dollar tomorrow depends on interest rates (Table 6–1). If someone offered to pay $107 one year from now, assuming the available rate of interest is 7%, then the present value of that payment would be $100.

Present Value

The formula for the present value of a single payment is

$$P = F/(1+i)^n$$

where

F = *the future value of a sum*
P = *the principal, or the present value of a sum*
i = *the rate of interest or discount rate*
n = *the number of time periods*
$1/(1+i)^n$ = *the discount factor*

Part of this formula, $[1/(1+i)^n]$ is referred to as the *discount factor*. It is multiplied by the future payment to arrive at its present value. F is said to be discounted by that factor. This is why the terms discount rate and interest rate are used interchangeably.

Assume that after five years a payment of $1,000 will be made. This is illustrated in Figure 6–2. The present value of that payment discounted at 10% for five years is equal to:

$$P = 1000/(1+.10)^5$$
$$P = 1000/1.6105$$
$$P = \$621$$

The analysis of a stream of payments or cash flows is based on discounting each future payment F, back to the present, hence present value.

The financial analysis of an oil company or even a single oil well is based on the present value of the expected stream of cash flow. These payments come in regularly, not just once a year. Because it is easier to make estimates based on annual figures, midyear discounting is normally used to emulate the nearly continuous income stream.

The formula for present value using midyear discounting is:

$$P = F/(1+i)^{n-.5}$$

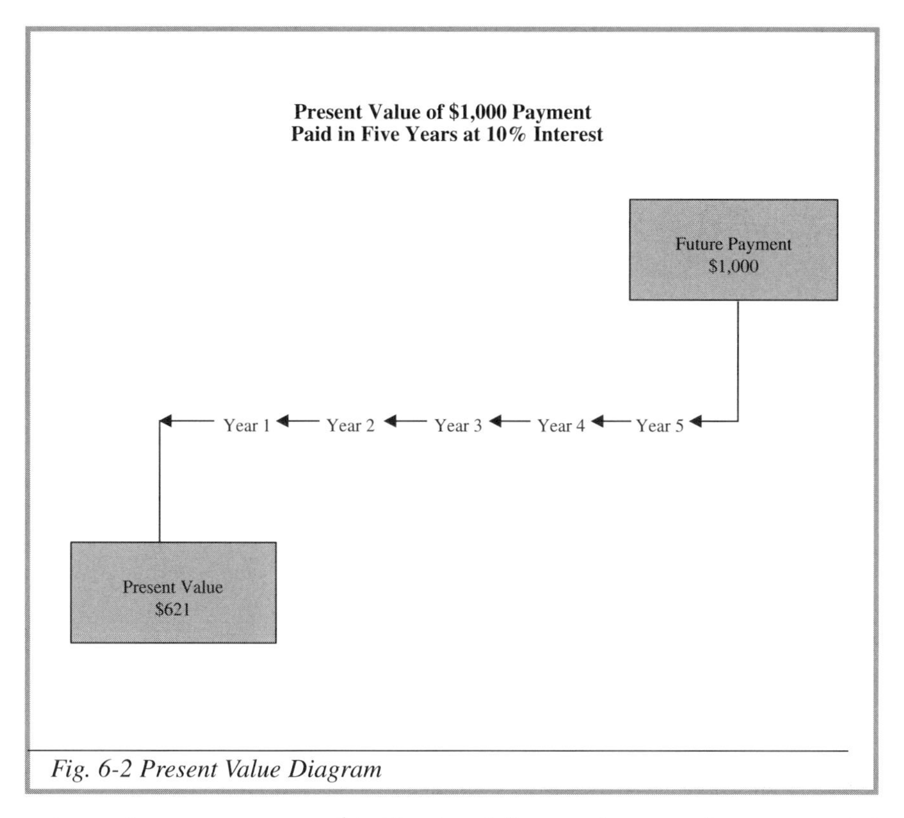

Fig. 6-2 Present Value Diagram

Assume that a company expects $1 million in cash flow over the course of a one-year period starting two years from now. The present value assuming a rate of interest of 10% would be:

$$P = \$1 \text{ million} / (1.1)^{3-5}$$
$$P = \$788 \text{ thousand}$$

Here the discount rate is 10%, and the discount factor is:

$$\text{Discount Factor} = 1/(1.1)^{3-5}$$
$$= 1/1.269$$
$$= .788$$

Although many spreadsheet programs default to end-of-year discounting. Most evaluation work uses midyear discounting.

Present Value Tables

Present value tables are provided in Appendix 6 and 7 for single payments or annuities. For example, assume a five-year cash flow stream that starts at $10,000 the first year and is expected to decline at a rate of 10% per year. The present value can be estimated by using the discount factors from Appendix 6. What would be the present value discounted at 15%? The example below shows the calculated present value of the declining cash flow using a midyear discount rate of 15%.

Year (n)	Declining Cash Flow (F)	Mid-Year Discount Factor (i=15%)	Present Value (P)
1	$10,000	.933	$9,330
2	9,000	.811	7,299
3	8,100	.705	5,710
4	7,290	.613	4,469
5	6,561	.533	3,497
			$30,305

The present value of the cash flow stream is $30,305.

Appendix 7 shows the present value for a series of equal payments--an annuity. For example, a five-year stream of cash flow discounted at 15% would have a present value of 3.595 times the annual payment. Thus, a cash flow stream of $10,000 per year for five years would have a present value (discounted at a rate of 15%) of $35,950.

Internal Rate of Return

Much of the business of financial analysis is the determination of present value based on a specific discount rate. Sometimes the objective is to essentially work backwards and calculate the discount rate. What discount factor would yield a present value of $25,000 for an annuity of $10,000 per year for five years? This would be a natural question if the annuity was for sale for $25,000. The answer to that question would be called the internal rate of return (IRR). The IRR is the discount rate at which the present value of a cash flow stream from an investment would equal the cost of the investment.

Corporations usually establish specific investment criteria and the choice of what discount rate to use to evaluate investment opportunities is important. The "required rate of return" is common language for choice of a specific IRR as is the concept of a *hurdle rate*. In the language of investment analysis a project may be required to exceed a specific *hurdle rate* also referred to as the *corporate investment hurdle rate*.

Assume that an investor was interested in buying a business that would provide $10,000 per year for five years. Income is relatively continuous, so a midyear discount rate is used—for this example the investor uses a discount rate (or a hurdle rate) of 20%.

Year (n)	Cash Flow (F)	Mid-Year Discount Factor (i=20%)	Present Value (P)
1	$10,000	.913	$9,130
2	10,000	.761	7,610
3	10,000	.634	6,340
4	10,000	.528	5,280
5	10,000	.440	4,400
			$32,760

The present value of the income stream could also have been estimated by using the table in Appendix 7. The multiplication factor for a five year stream of income discounted at 20% (midyear) is 3.276. If the investor paid $32,760 for the business, then the internal rate of return on that investment would be 20%. If he paid less than that, the IRR would be greater. A payment greater than $32,760 would yield an IRR of less than 20%.

What if he paid $35,000 for the enterprise? The IRR then would be whatever discount rate (i) it would take to produce a present value of $35,000. This calculation requires a trial-and-error procedure. The IRR will be less than 20%, but by how much? Computers try alternate discount rates to close in on the answer. This is called an *iterative approach*. The answer is approximately 16.4%. That is, the IRR of the investment would be 16.4% if the purchase price is $35,000. Another way of putting it would be that the present value of the five-year stream of $10,000 payments discounted at 16.4% is $35,000.

RISK ANALYSIS

One of the most interesting and important aspects of investment theory is the subject of risk. The definition of risk is slightly confusing because it is a bit difficult to define in such a way as to make all people happy. *Risk analysis* is the business of evaluating possible outcomes and uncertainties and characterizing in a reasonable fashion those uncertainties so that good business decisions can be made. In order to do this, proper analysis should (in the language of risk analysis) preserve the uncertainty.

Two people may evaluate a drilling deal and their evaluation may look like this:

1st Analyst: *It looks like a good deal, we should participate. The prospect could hold 10 million barrels. The odds look good, the drilling costs are not too high, and the terms are good.*

2nd Analyst: *Our geologists estimate a probability of success of over 30% which is substantially higher than breakeven success probability (SP) of 18%. The most likely reserve estimate is around 10 million barrels, with a range of from 5-18 million barrels. The regional success rate is close to 25%. The expected reserves on this prospect are 3 million barrels and the expected monetary value (discounted at 15%) at a 30% chance of success is over $1 million. Dry hole costs are less than $1.5 million. The prospect meets our basic investment criteria and ranked with the other good prospects is No. 3 behind Prospect X and Prospect F.*

This example may appear to be a bit drastic, yet it captures the diversity of perspective fairly well. The degree of sophistication can range from rank amateur to the lofty reaches of analytical sophistication. However, there is not a perfect correlation between sophistication and accuracy. Gut instinct or intuition, backed up by 30 years of experience, can seem very unsophisticated at times but can be very valuable.

It is important to communicate at all levels, as hard as that may be at times. And the place to start is with the definition of risk—yet even that is not as simple as it may seem.

Decision analysis, sometimes called 'risk analysis,' is the discipline for helping decision makers choose wisely under conditions of uncertainty.
–John R. Schuyler, *Decision Analysis in Projects*™ (pp. 3)

..And, indeed, an often confusing intellectual tussle ensues over the difference between risks and uncertainties. What do we have here? A risk or an uncertainty?
–Dr. John Lohrenz, *Certain Uncertainties* (pp. 3)

Many people equate risk with uncertainty.

We will consider the words 'risk' and 'uncertainty' to be synonymous"
-Paul Newendorp, *Decision Analysis for Petroleum Exploration* (pp. 59)

"Risk vs. Uncertainty"

Some do not equate risk with uncertainty.

Many people equate risk to uncertainty. In this chapter we will make a distinction between the two. We shall always use risk to mean an opportunity for loss. The term uncertainty shall apply to factors where the outcome is not certain but where the opportunity for loss is not as apparent as in risk.
–Robert E. Megill, *Exploration Economics* (pp. 140)

RISK. A measurable possibility of losing or not gaining value. Risk is differentiated from uncertainty, which is not measurable.
–Barron's Dictionary of Finance and Investment Terms (pp. 348)

Some relate risk with uncertainty.

Risk analysts start by dividing hazards into two parts: exposure and effect. . . . The study of exposure and effects is fraught with uncertainty. Indeed, uncertainty is at the heart of the definition of risk. In many cases, the risk may be well understood in a statistical sense but still be uncertain at the level of individual events. Insurance companies cannot predict whether any single driver will be killed or injured in an accident, even though they can estimate the annual number of crash-related deaths and injuries in the U. S. with considerable precision.
–M. Granger Morgan, "Risk Analysis and Management," *Scientific American*, July 1993 (pp. 32)

The term 'risk' is well understood and almost universally misused, not excepting the following discussion. Risk has come to mean the chance of failure or loss, but it is actually associated with and defined by the dispersion of the possible outcomes; that is, risk is a measure of the degree of uncertainty and does not necessarily reflect a high probability of a bad outcome. Risk exposure reflects the penalty for failure.
—I. Field Roebuck, Jr., *Economic Analysis of Petroleum Ventures* (pp.11)

The terms' risk' and 'uncertainty' are frequently used almost interchangeably in everyday discussion. For our purposes this will be differentiated. Uncertainty will be used to characterize the fact that the eventual outcome of a decision or event is not precisely known, with the degree of uncertainty described by the probability that it will occur...

Risk, on the other hand, denotes that there is a possibility of incurring economic loss or reduced value. High risk ventures are ones with a chance of a large loss, even if the probability of such an occurrence is small. It is possible for a project to be highly uncertain but have a low risk, if failure would be inconsequential.
—Fraser H. Allen, Richard D. Seba, *Economics of Worldwide Petroleum Production* (pp. 191)

Another perspective. Two investors are contemplating similar investments. The first investor is guaranteed a rate of return of 12%. The second investor is guaranteed at least a 12% rate of return, but has the potential to receive as high as 18%. Does the second investor's uncertainty regarding his potential ultimate ROR make his investment more risky? No.

Field Size Estimates vs. Deliverability Estimates

One of the most important aspects of risk analysis in the oil industry deals with the uncertainty of ultimate recovery. Unfortunately, too much attention is often placed upon recoverable reserves when, in fact, deliverability can be more critical. Because of the time value of money, reserves produced after five to ten years have substantially less impact from a financial point of view. Table 6–2 emphasizes this point.

A typical production profile is shown with a decline rate of 12%. A mid-year discount rate of 12% is used for present value weighting purposes. Notice that the first 15 million barrels (first year of production) has a weighted value of 22.5% compared to the last 15 million barrels (Years 10–13) which contribute only 7% of the weighted value. Notice too, that over 50% of the weighted value of the reserves accumulates within the first three years of production. This kind

Total Reserves = 101 Million Barrels
Production 1st Year = 15 Million Barrels
Decline Rate = 12%
Discount Rate = 12%

Year	Production (MMBBLS)	Midyear Discount Factor	Present Value Weighted Reserves	Weighted %	Cumulative
1	15.0	.945	14.2	22.5%	22.5%
2	13.2	.844	11.1	17.7	40.2
3	11.6	.753	8.8	13.9	54.1
4	10.2	.673	6.9	10.9	65.0
5	9.0	.601	5.4	8.6	73.6
6	7.9	.536	4.2	6.7	80.4
7	7.0	.479	3.3	5.3	85.7
8	6.1	.427	2.6	4.2	89.8
9	5.4	.382	2.1	3.3	93.1
10	4.7	.341	1.6	2.6	95.7
11	4.2	.272	1.1	1.8	97.5
12	3.7	.243	.9	1.4	98.9
13	3.2	.217	.7	1.1	100.0
	101.2		62.9		

Table 6–2 The Importance of Deliverability vs. Total Reserves

of timing relationship must be kept in mind when reviewing cash flow projections based upon reserve estimates. This is also why it is often pointed out that 85% of the Saudi's oil reserves have no value from a present value point of view.

EXPECTED VALUE THEORY

The decision-making process associated with placing money on a drilling location can range from pure gut level experience to sophisticated analytical methodology. Regardless of the level of sophistication, the experience factor and the gut level choices are of great importance. Even for the relatively unsophisticated decision maker, the process is much the same. At the gut level, investors either consciously or subconsciously weigh in their minds the balance between risk and reward. The only difference is that some people put a name to that approach. It is

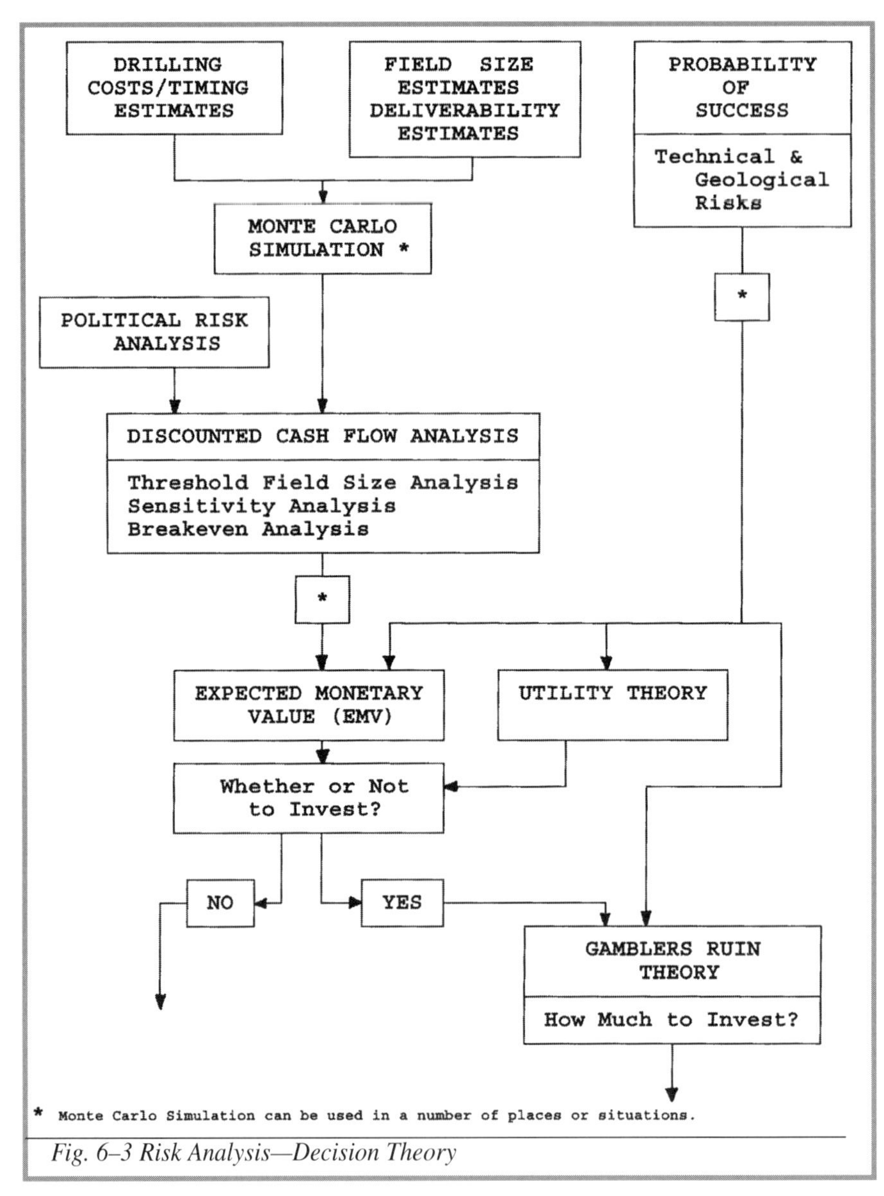

Fig. 6–3 Risk Analysis—Decision Theory

called *expected value theory* (EVT). Figure 6–3 shows where EVT and other analyses fit into the decision theory scenario.

The use of expected value theory—also often referred to as the *expected monetary value approach* (EMV)—became more common throughout the 1980s as a formal tool of decision makers. Prior to that there were many who informally weighed the risks and rewards of a possible drilling deal and then made their decision. The decision of whether or not to drill is one of the best examples of the use of this tool. The EMV weighs risk capital and the chance of losing it with the potential reward and the probability of achieving that reward. The EMV formula is:

$$EMV = (Reward \times SP) - [Risk\ capital \times (1 - SP)]$$

where

Reward	=	*Present value of possible successful drilling results*
Risk Capital	=	*Dry-hole costs and other up-front costs*
SP	=	*Success probability*

This formula provides the cornerstone of risk analysis. The rule is that if EMV is positive, then the risk-weighted reward outweighs the risk-weighted cost of failure.

Reward

In a drilling project, the reward is represented by the discounted present value of estimated successful drilling results. The rate at which the cash flow projection is discounted should be a discount rate that would represent an acceptable rate of return to the company. Sometimes a hurdle rate is established by corporate financial management for discounting purposes. The issue is based upon the corporate cost of capital and/or the perceived market rate of interest for such investments. The decision to go forward with a project therefore could be made for a project that had any positive expected value.

Risk Capital

Risk capital is usually the costs associated with the drilling of an exploratory well, seismic data acquisition and processing, site preparation, and dry hole costs, etc. The importance of the risk capital cannot be overemphasized. The relationship between risk capital and the present value of a successful venture is a function of success probability. In an area where the probability of success is as high as 20%, the risk dollars must be offset by a factor of at least 5 to 1. This is the essence of expected value theory.

Success Probability

Estimation of the probability of occurrence of either a dry hole or some form of discovery is one of the more difficult aspects of the evaluation process. Many people are very uncomfortable making such estimates. There is perhaps more gut instinct required at this stage than any other. Some analysts will run a cash flow to determine possible financial results to three decimal places and then go pale at the thought of estimating success probability (SP). Sometimes it helps to simply calculate a break-even success rate, and then determine how that compares with estimates of the probability of success (also called *chance factor*). If break-even SP is around 20% in a region where success ratios are over 25%, then the matter becomes less complicated.

Figure 6–4 illustrates the relationship between the key variables in EMV theory--a case is outlined in which dry-hole costs for a particular drilling deal are $30 million. If the well is dry, the investment is lost. If the well turns out as hoped, then the present value of the project discounted at 15% is estimated at $110 million.

If management believed that the probability of success were greater for the project they may be interested in investing. The EMV approach is the formal method of weighing the possible out-

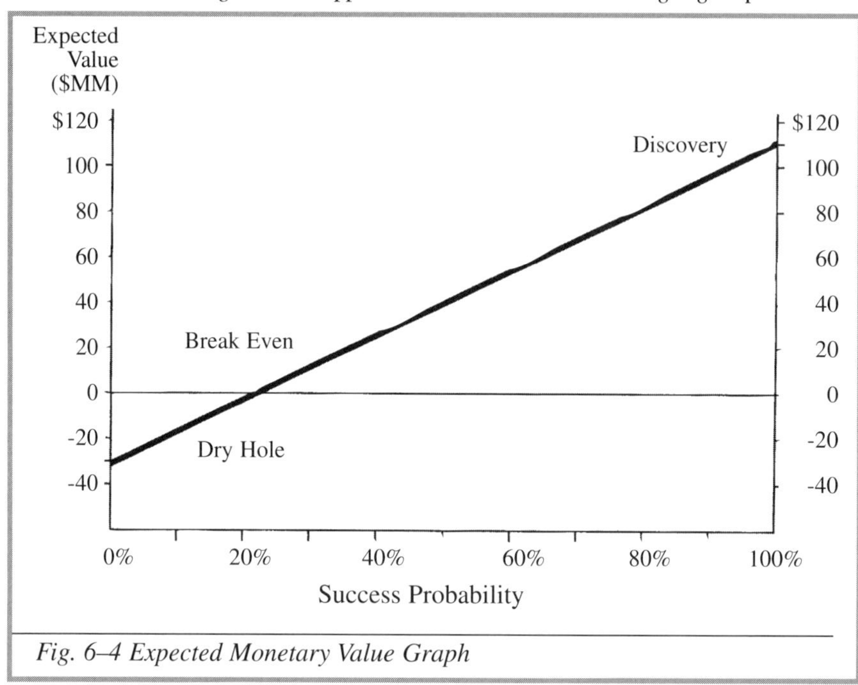

Fig. 6–4 Expected Monetary Value Graph

comes. Suppose management believed that the probability of success were around 30%. Table 6–4 shows that the EMV then would be $12 million.

The drilling would either result in a total loss or a success but on the average, this kind of play or deal would enhance the corporate wealth by around $12 million if enough such similar oppor-

Possible Outcome	Present Value ($MM)	Probability (%)	Expected Monetary Value ($MM)
Reward	$110	30%	$33
Dry Hole	-30	70%	-21
			$12

Table 6–3 Expected Monetary Value

tunities were available, and assuming the estimates of costs, rewards and probabilities are correct.

With these monetary risks and that kind of potential reward, management would not go near the project unless it estimated a probability of success for the well greater than 21%. This is the break-even success ratio which is determined either graphically or by setting EMV equal to zero in the expected value formula:

$$0 = (\text{Reward x SP}) - [\text{Risk Capital x } (1 - \text{SP})]$$

Solving for SP, the above formula converts to:

$$SP_{be} = \text{Risk Capital} \div (\text{Reward} + \text{Risk Capital})$$

where

EMV	=	0 (zero) by definition
Reward	=	$110 MM
Risk Capital	=	$30 MM
SP_{be}	=	Break-even Success Probability
SP_{be}	=	21.4%

The break-even success ratio is related to a concept called success capacity. Success capacity is the number of dry holes a successful well can carry. It is equal to the inverse of the break-even success ration minus 1:

$$\text{Success capacity} = (1/\text{break-even success ratio}) - 1$$

The success capacity in the example above is 3.7, (1/.214)–1.

Drill or Farm Out?

One of the most common examples of the use of this graphical representation of a two outcome EMV analysis is the comparison of whether or not to drill or to farm out. Suppose there were an opportunity to farm out half of this prospect rather than drilling alone. Assuming that a partner would be willing to incur the cost of the exploratory well, for 50% working interest. This is referred to as a *drill-to-earn* type of deal, and it is fairly common with numerous slight variations.

The farmout option requires no risk dollars from the perspective of the license holder in this example, but the present value of a potential discovery is half of what it would have been without farming out 50% of the working interest. The results depicted in Figure 6–5 illustrate that if management believed that the success probability is less than about 33%, then the best strategy would be to farm out the prospect. Below 33% success probability the expected value is always higher with the farmout strategy. Above that point the strategy to farmout has a lower EMV.

The potential partner's perspective is shown in Figure 6–6. Suddenly, the importance of the risk dollars becomes more clear. Now the breakeven success probability is closer to 40% for

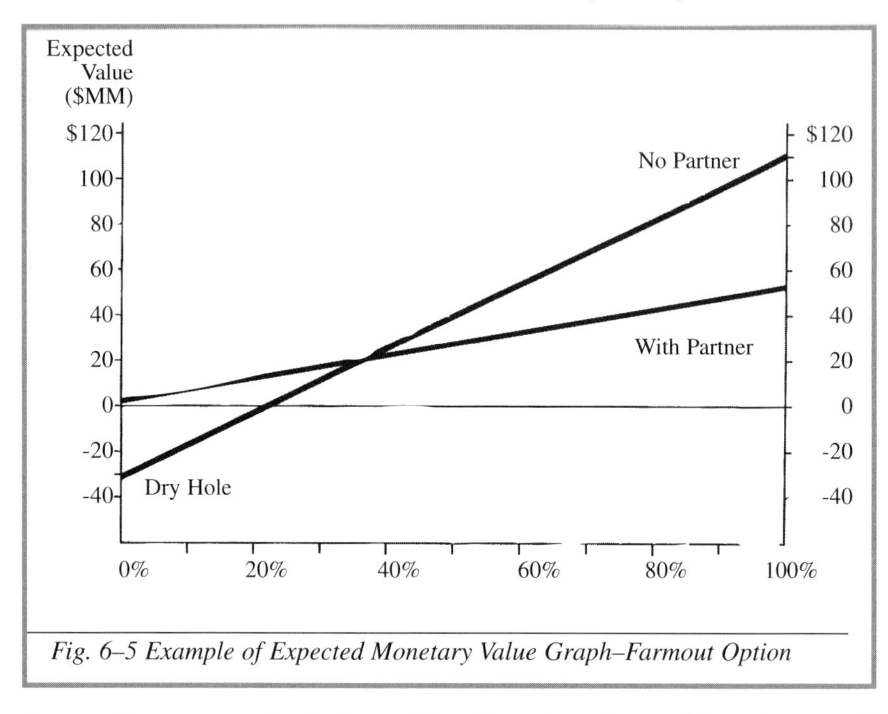

Fig. 6–5 Example of Expected Monetary Value Graph–Farmout Option

the potential partner, assuming that he would be willing to incur the cost of the exploratory well for 50% working interest.

There are six basic steps in the expected monetary value decision analysis as follows:

Expected Monetary Value Decision Analysis

1. Define possible outcomes or events and decision alternatives.
2. Estimate probability of occurrence of each outcome or event.
3. Calculate monetary value for each possible outcome. (Discounted cash flow analysis).
4. Multiply probability of occurrence with monetary value of each possible outcome.
5. Calculate algebraic sum of the expected values of all possible outcomes.
6. Select alternatives which maximize EMV.

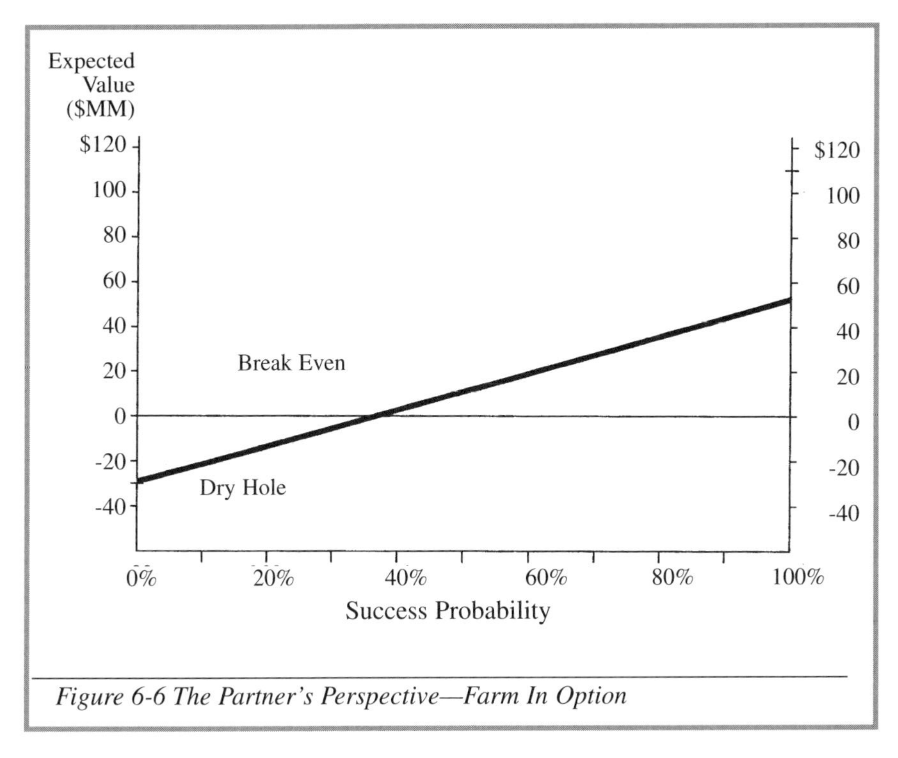

Figure 6-6 The Partner's Perspective—Farm In Option

DECISION TREE ANALYSIS

Decision tree analysis is pure EMV theory. The previous examples are sometimes described as two-outcome expected value models or two-outcome decision trees. The EMV approach can be logically applied to a number of scenarios. The probabilities of occurrence for each possible event must add up to 100%. An example is outlined that has four possible events that could result from the decision to drill. If the well is a dry hole, the potential loss is $15 million. A discovery could be worth from $20 million to $450 million in this model (Fig. 6-7).

The multiple-outcome decision tree in Figure 6–7 has been collapsed to a simple two-outcome tree (Figure 6–8).

The Value-of-Information Concept

One of the smaller, less frequently visited branches on the tree of risk analysis is the concept of the value-of-information. An example is used here to demonstrate how it can look. A two-outcome decision tree is used as an example. The following Table 6-4 summarizes what the expected results would look like depending upon whether or not a 3-D seismic survey is implemented. The example here assumes that with 3-D seismic data the success ratios improve.

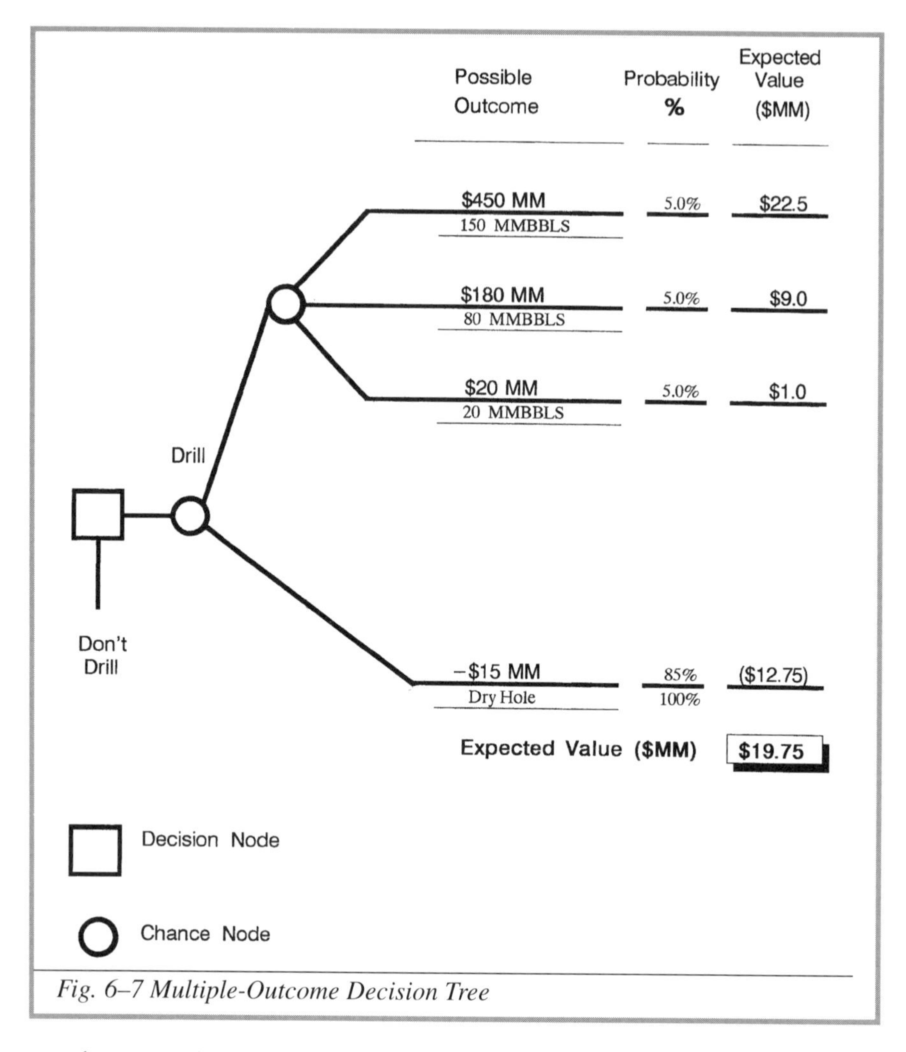

Fig. 6–7 Multiple-Outcome Decision Tree

Without 3-D in this example, the success probability is 30%. But, suppose that where drilling locations have been identified with 3-D programs the success rates are better. In this example, five percentage points better. The expected value improves by $7.25 million. But notice in this example, the cost of the 3-D survey has not yet been accounted for. This way, it is easy to see that the cost of the 3-D survey must be kept below $7.25 million or the drilling decision is better off without the value of that information.

Diversification of Risk

Hand-in-hand with the decision to participate in a drilling venture or just about any investment, is the question of how much exposure the company should incur. This is also known as

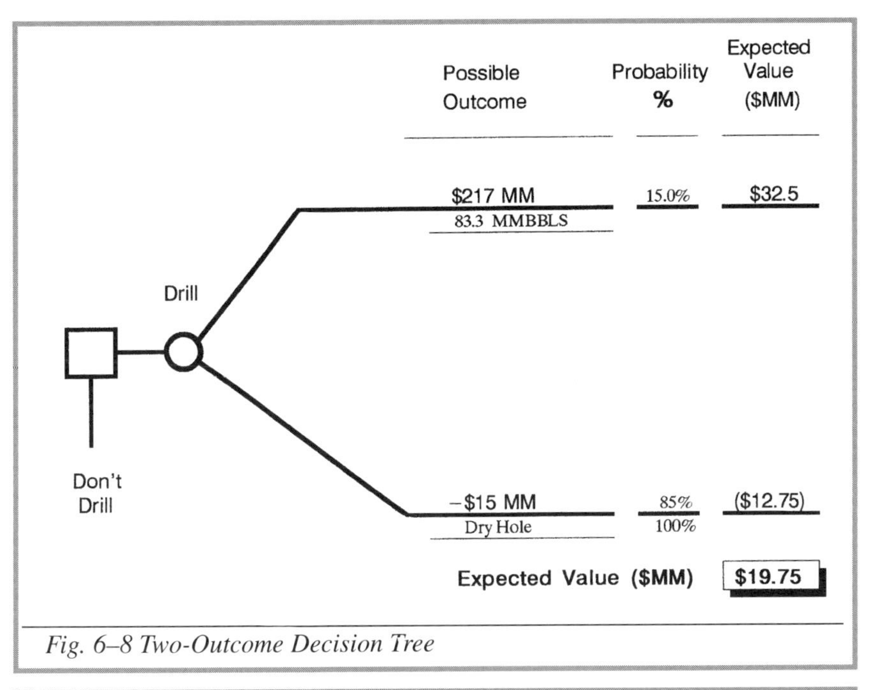

Fig. 6–8 Two-Outcome Decision Tree

Two-Outcome Model Without 3-D Seismic

Possible Outcome	Present Value ($MM)	Probability (%)	Expected Monetary Value ($M)
Reward	$120	20%	$ 24.00
Dry Hole	- 25	80%	- 20.00
			$4.00

Two-Outcome Model With 3-D Seismic

Possible Outcome	Present Value ($M)	Probability (%)	Expected Monetary Value ($M)
Reward	$120	25%	$ 30.00
Dry Hole	- 25	75%	- 18.75
			$ 11.25

Table 6–4 Value of Information Concept

the eggs-in-the-basket branch of investment philosophy. The general rule is that there shouldn't be too many in any particular basket. The difference between a gambler and an investment strategist in many drilling deals lies in the degree of exposure (Fig. 6–9).

Expected Monetary Value Formula

```
EMV  =  (Reward * SP) - [Risk capital * (1-SP)]
EMV  =  ($110MM * .30) - [$30MM * (.70)]
EMV  =  $33MM - $21MM
EMV  =  $12MM
```

Expected Monetary Value Graph

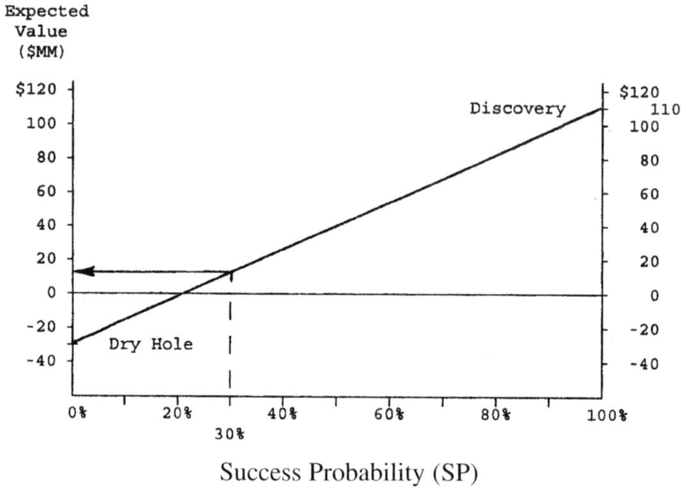

Expected Value ($MM)
$120
100
80
60
40
20
0
-20
-40
Discovery
Dry Hole
$120
110
100
80
60
40
20
0
-20
-40
0%
20%
30%
40%
60%
80%
100%

Success Probability (SP)

Two-Outcome Decision Tree

Drill
Do Not Drill

Possible Outcome		SP	EMV
Discovery	$110MM	30%	$33MM
Dry Hole	-$30MM	70%	$21MM
	Expected Value	100%	$12MM

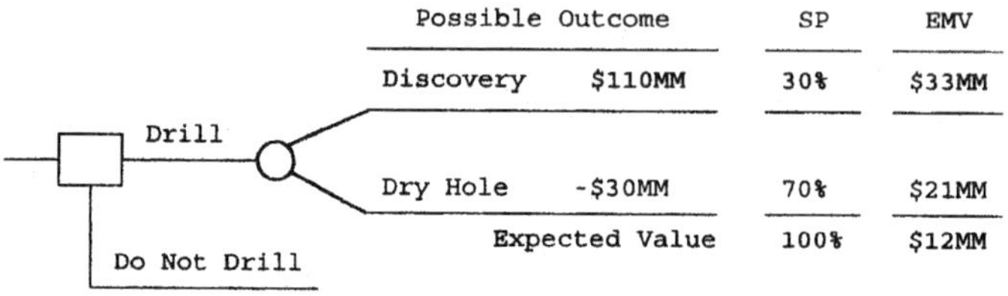

Tabular Format Two-Outcome Decision Tree

Possible Outcome	NPV ($MM)	Probability (%)	EMV ($MM)
Reward	$110	30%	$33
Dry Hole	-30	70%	- 21
		EMV	$ 12

Fig. 6–9 Four Valuation Methods with The Same Result

GAMBLER'S RUIN THEORY

Gambler's Ruin occurs when a risk-taker with a limited amount of funds goes broke through a continuous string of failures. With any kind of drilling budget there is such a risk, but diversification can minimize this risk. The concept of Gambler's Ruin flows from the same estimates of risk used in expected value decision theory. The estimate of the probability of success is used to help determine the maximum level of capital exposure under a specific confidence level criteria. The objective for management is to stay in the drilling game long enough for the odds to work as they should.

Another way of looking at it would be to ask the question, "What would be a statistically significant sampling size (at a confidence level of 90%) given the nature of the play?" The concept of Gambler's Ruin is used for this. To avoid a successive string of failures that would exhaust a drilling budget for example, at least one success must be achieved. The Gambler's Ruin algorithms are outlined as follows in the next section.

The probability of at least one success = *1 – The probability of all failures*

Assume that a confidence level of 95% is desired. How many wells must be drilled to be 95% confident that we will have at least one success? Too often the assumption is made that with a one in five chance of success with five wells that at least one will be successful. There is a 33% chance this will not happen.

GAMBLER'S RUIN ALGORITHMS

$$.95 = 1 - (1 - sp)n$$

where

.95	=	*Desired confidence level*
sp	=	*Success probability*
(1 - sp)	=	*Probability of failure*
n	=	*Number of trials (exploratory wells)*

$$n = \frac{\log (0.01)}{\log (1 - sp)}$$

where

0.01	=	*1 – Desired confidence level* *Probability of all failures*
sp	=	*Success probability*
1 – sp	=	*Probability of failure*
n	=	*Number of trials (exploratory wells)*

The results of this formula are shown in Figure 6–10. If five wells are drilled with probability of success of 20%, there is a 67.2% confidence level of at least one successful well. All of the options have been incorporated in the Gambler's Ruin calculation.

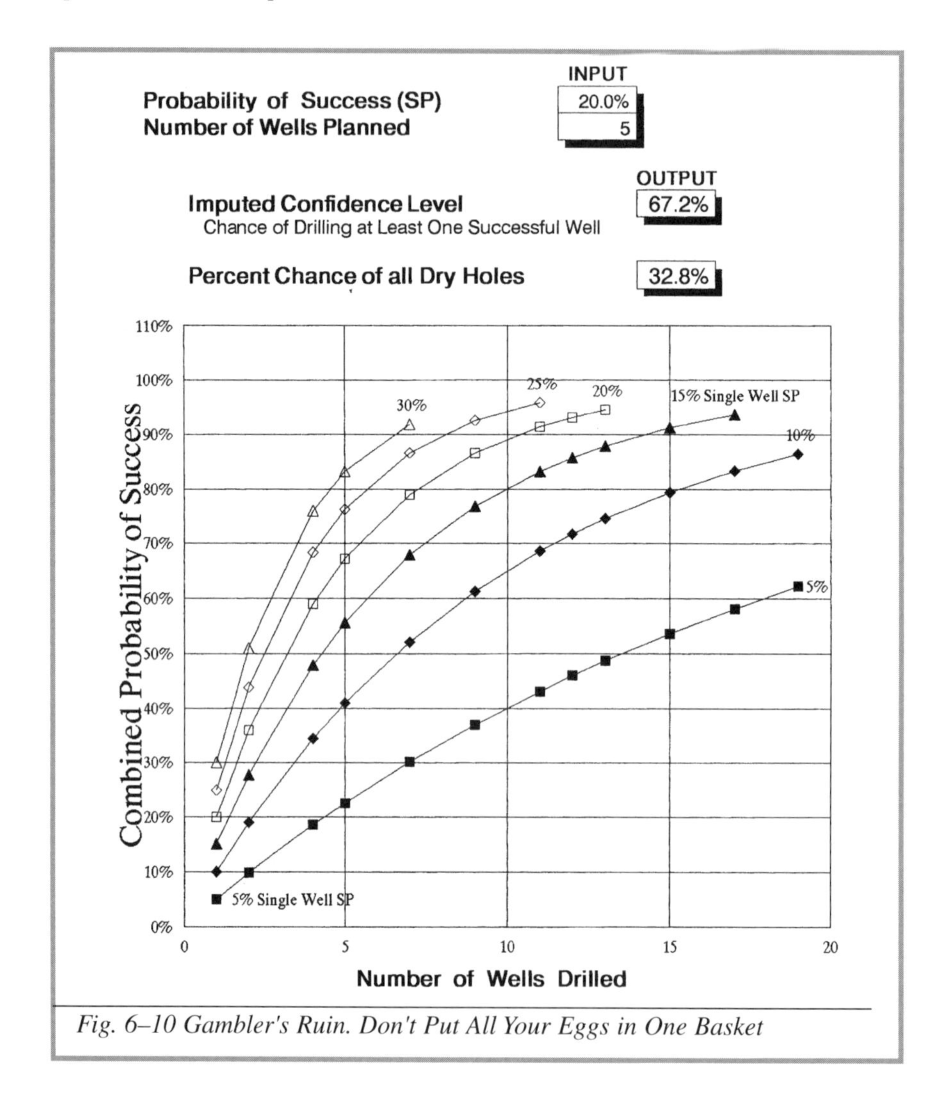

Fig. 6–10 Gambler's Ruin. Don't Put All Your Eggs in One Basket

Estimating Probabilities

With all of the expected value approaches, the cornerstones of the analytical process are founded on the present value of projections of cash flow and the estimation of success probability. Estimating the probability of success or failure is often part science and part guesstimation.

Sometimes, regardless of the degree of sophistication involved, the whole process boils down to a gut feel about a particular project. This approach is wonderful if the gut feeling is founded on an understanding of the implications of EMV theory, as well as years of experience. The concept and use of success probability (SP) estimates is not as simple as one might hope.

When viewing the investment decision, the estimate of SP must be based upon experience and available knowledge. Often in new frontiers and exploration plays, the available information can be scarce. Regardless of the experience involved, the decision makers must know the basics of the odds out there in the oil patch. It helps to understand some of the things that influence SP and the first step along this road is an understanding of the difference between commercial and technical success.

Commercial and Technical Success

When a well actually locates a measurable quantity of either oil or gas, the well is often classed as a technical success. Whether or not the well will ever be worth the money invested, it is a measure of commercial success. In many places the conventional wisdom or the general experience in the area will provide some indications of success rates for both exploratory and development drilling. These rates unfortunately are often the technical success rates not the commercial rates. Technical success rates are always higher than commercial success rates. The investment decision really should be based upon an understanding of the difference.

Another way of looking at this issue is: technical success ratio is the chance of finding hydrocarbons. The commercial success ratio is the chance of finding something large enough to develop. The difference, therefore, between technical and commercial success is the development field size threshold.

Some areas or types of plays are strongly characterized by the difference between commercial success and technical success and by the degree of success and success rates, see Table 6-5.

Sensitivity Analysis

Sensitivity analysis is done by systematically varying each parameter and observing the change in project IRR, NPV, payout, and ROI, etc. But, the results are nearly always the same. The most critical parameters are product prices and capital costs. Furthermore, these two parameters are the ones that are most likely to vary substantially. The economic outcome of a potential development is usually much less sensitive to other parameters and those parameters are usually more predictable and less prone to variation.

Simple Example

A	Source/Timing	70%	Estimate
B	Reservoir	90%	Estimate
C	Trap/Seal	60%	Estimate
D	Structure	80%	Estimate
E	Probability of Hydrocarbons	30%	A*B*C*D Sometimes called Geologic Probability

More Complex Example

A	Source	95%	Estimate
B	Trap	70%	Estimate
C	Ability to Locate	60%	Estimate
D	Reservoir Quality Rock Exists	90%	Estimate
E	Timely Migration	75%	Estimate
F	Other	100%	Estimate
G	Probability of Hydrocarbons	27%	A*B*C*D*E*F Sometimes called Geologic Probability
H	Probability of Oil	60%	Estimate - Assuming hydrocarbons are found
I	Oil SP	16%	G*H
J	Gas SP	11%	G* (1 – H)
K	P of oil exceeding threshold size	80%	Estimated probability of exceeding minimum reserves given a discovery. Threshold field size.
L	P of gas exceeding threshold size	30%	Estimated probability of exceeding minimum reserves given a discovery. Threshold field size.
M	P of commercial oil discovery	12.8%	I*K oil only
N	P of commercial gas discovery	3.3%	J*L gas only
O	P of commercial discovery	16.3%	M+N Oil or Gas

P = Probability SP = Success Probability

(In these examples the variables are considered to be independent of each other)

Table 6–5 Two Success Probability Estimation Methodologies

The factors that are often evaluated to determine the sensitivity of project economics are:

- Product Prices
- Product Price Escalation
- Capital Costs
- Operating Costs
- Timing of Production Startup

- Expected Ultimate Recoverable Reserves
- Reserves Deliverability
- Success Ratios

Various techniques are used to summarize or characterize project economics sensitivity. Spider diagrams are relatively rare but demonstrate graphically the financial results of different scenarios (Fig. 6–11). Generally, the curves with the greatest slope indicate the variables that have the most impact on the financial aspects of the project. However, the graph can be misleading. Sometimes the likelihood of a particular parameter changing by perhaps 20% is so slim as to render it unworthy of consideration, yet all the other parameters are varied by plus or minus 15 to 20 percentage points. Tornado diagrams are used in much the same way.

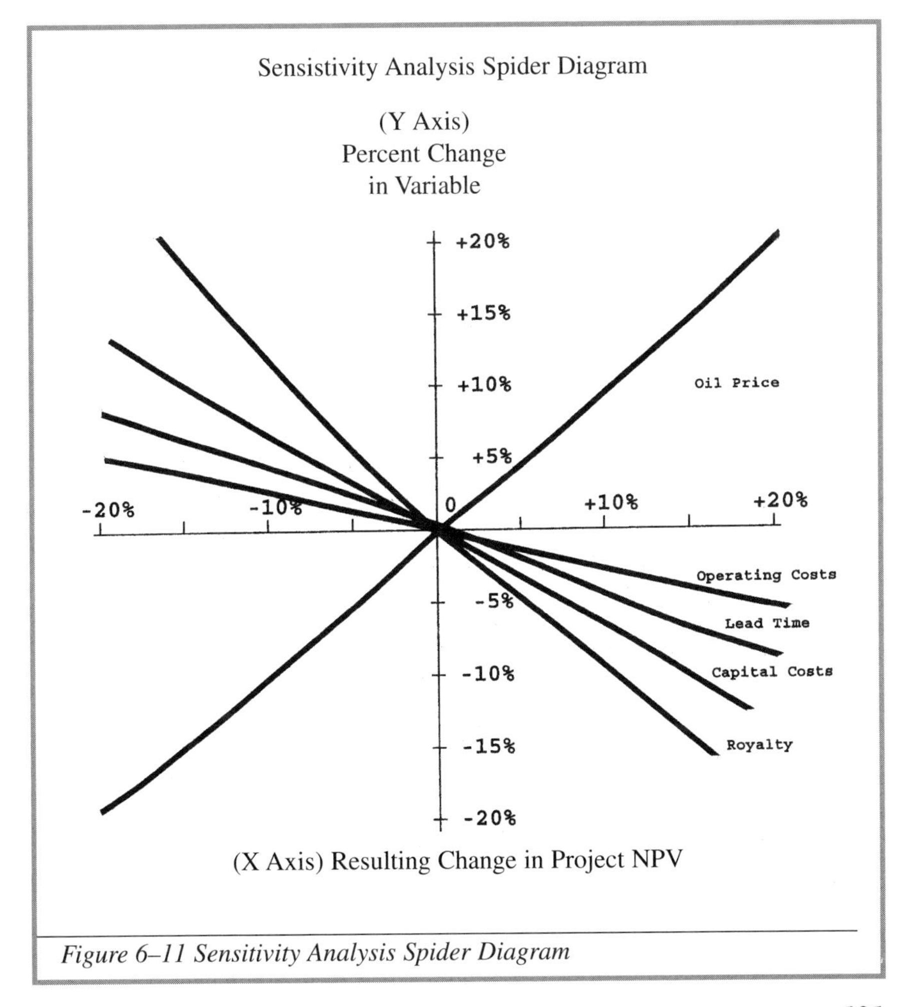

Figure 6–11 Sensitivity Analysis Spider Diagram

COMMON PITFALLS IN RISK ANALYSIS

Using discount rate to "risk" a project. One of the more common inappropriate techniques to account for project risk is to use a higher discount rate for higher-risk projects. This is consistent with one aspect of modern portfolio theory and the capital asset pricing model (CAPM) that dictates a higher targeted rate of return for higher risk stocks or investments. However, it is seldom acceptable in exploration economics. While the CAPM uses beta to quantify risk or uncertainty, expected value theory uses estimates of the probability of success to characterize, quantify, and account for most of the risk.

Use of inappropriate estimate of success probability. Too often this factor is over-exaggerated in order to sell management on a particular play concept or prospect. One example that provides a good yardstick is the Gulf of Mexico. This province is characterized to a large extent by the fact that the seismic data quality is quite good. Furthermore, the province is relatively mature and well understood by international standards with substantial drilling activity and history. Wildcat drilling during the 1970s and 1980s was more than 200 wells per year. The cumulative wildcat success ratio was around 15% during the 1970s. By 1990, it had crept up to more than 20% due partially to the increased use of 3-D seismic data. Development well success ratios are on the order of 70-75% in this province. Some fields have very few associated dry delineation/development wells while others have more.

Assuming "development drilling" is risk free! In some areas of development drilling is nearly as risky as exploration drilling. This is an unusual case but to assume better than 90% chance of success for a series of development wells would be a mistake in most provinces.

Assuming that field size distributions are always either normally distributed or lognormally distributed. Some plays and types of reservoir exhibit strong log normal distributions in regard to reserve size (EUR). However, other reservoirs or play types can be treated as though they are normally distributed. The normal distribution is easier to evaluate. Log normal distribution needs close scrutiny and multi-outcome modeling. Usually the two-outcome EMV approach is sufficient for plays with an expected normal distribution. However, for log normal distributions even a three-outcome modeling of a possible discovery may not be sufficient.

UTILITY THEORY

Human and corporate behaviors manage to carry on whether or not people know or care that an exotic name is applied to their actions. Utility theory is also known as *preference theory*.

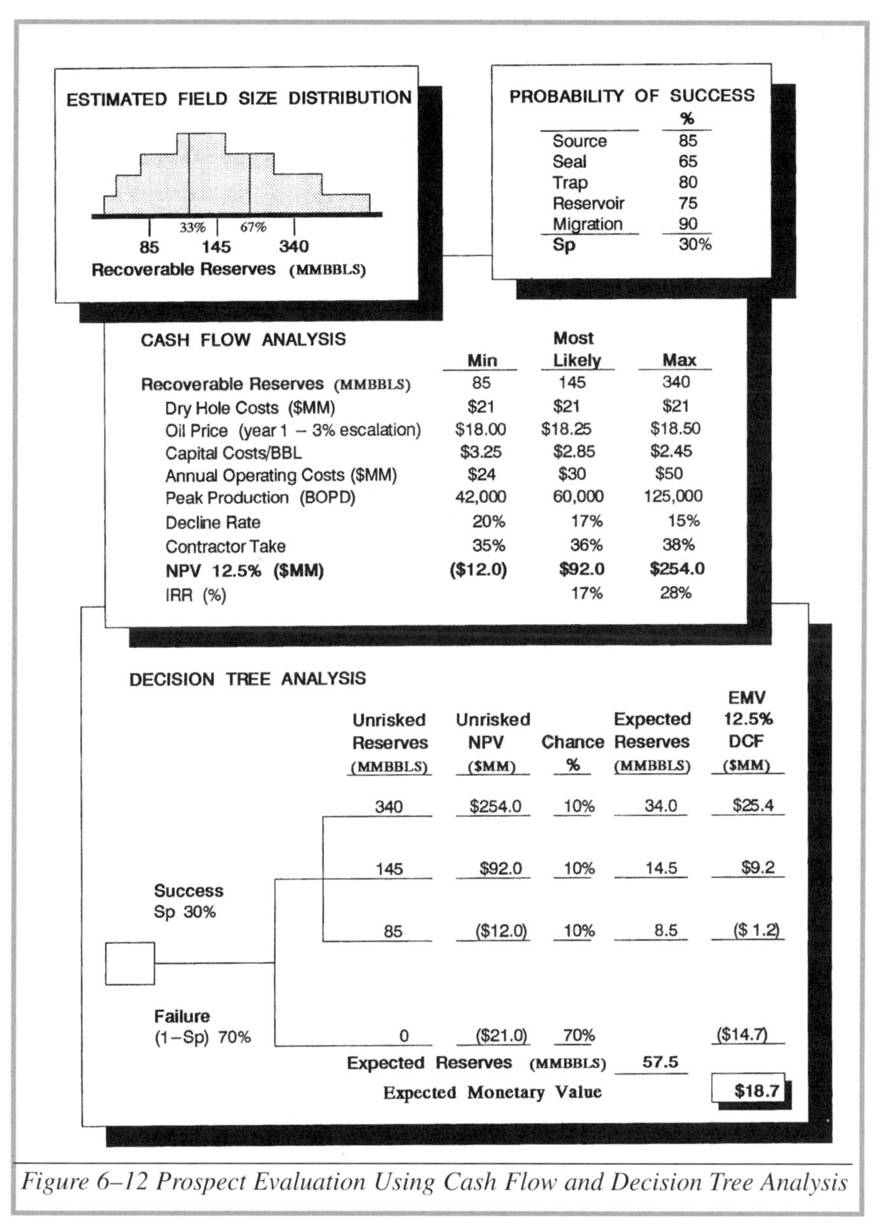

Figure 6–12 Prospect Evaluation Using Cash Flow and Decision Tree Analysis

It describes to a large extent why people will happily stick a quarter into a slot machine in Las Vegas even though the odds are squarely against them. The expected value of that sort of action is always negative–always. This is called gambling.

But quarters have almost no utility. When it comes to risking a few million dollars on an exploratory well, even expected value theory is not enough. Expected monetary value theory explains what people should do and what the boundary conditions are. It does not explain behavior. If the expected value of a potential investment opportunity is positive, then it is worthy of consideration. Just how positive must an expected value be? The standard industry two-outcome EMV model is used once again here in Figure 6-13, to illustrate the essence of utility theory.

The risk capital is $15 million. The possible reward in this two-outcome model has a value of just over $100 million. These points define the EMV curve. Staying below the curve results in positive expected values. For example, the expected value is equal to $20 million with an estimated probability of success of 30%. If management were bidding on this project, the bonus bid would have to be less than $20 million.

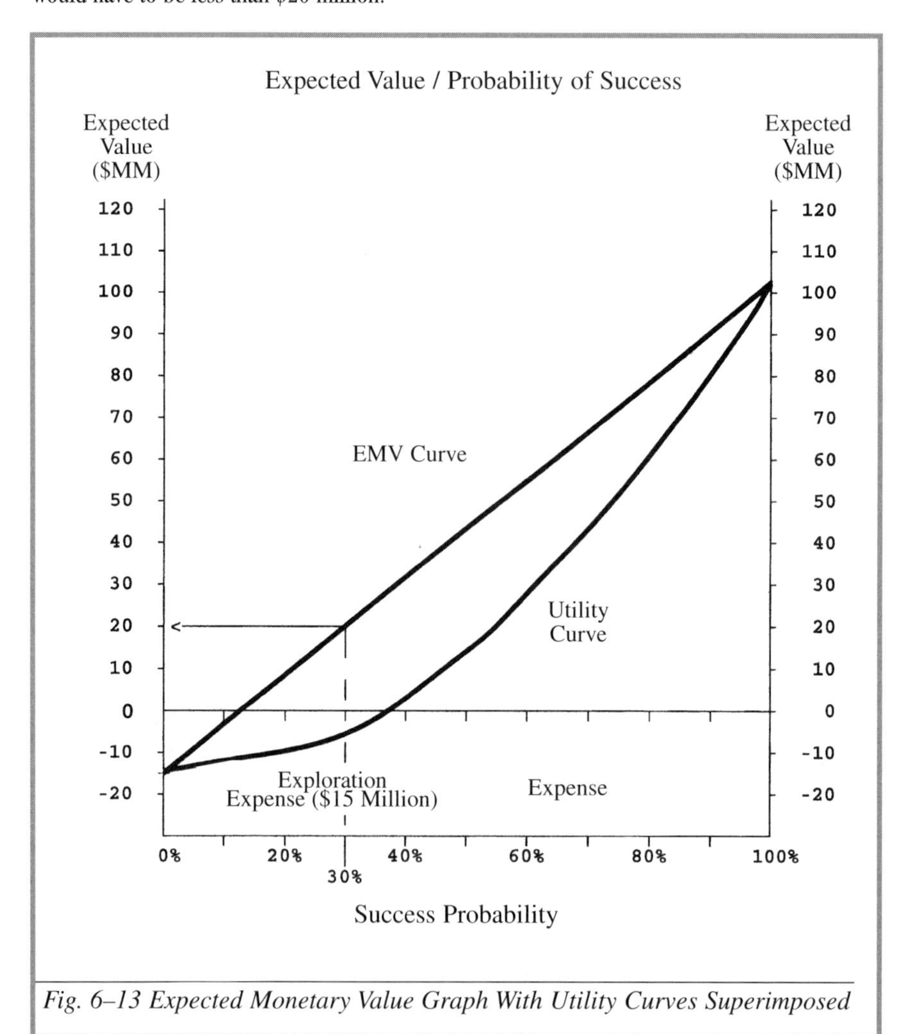

Fig. 6–13 Expected Monetary Value Graph With Utility Curves Superimposed

The EMV breakeven success ratio is close to 13%, but the utility breakeven success ratio is over 25%. Utility curves and EMV curves approach each other at the end points. If management was convinced that the $15 million well had zero percent chance of success, then it would be worth a negative $15 million. No company would drill such a well unless someone paid the company $15 million to do it. Drilling contractors do it all the time. On the other hand, if management were convinced that this drilling opportunity had a 99% or 100% chance of success, the expected value and utility values again converge. It would be like buying production or proved undeveloped reserves. It is in the middle ground where the curves diverge. Companies have different risk profiles or levels of *risk aversion*. Under the same set of assumptions, the EMV curves would look the same but the utility curves would not. Determining a company's utility curve with a drilling venture like this is complicated and becomes academic—consequently, it is not done often. Sometimes the whole subject is simply marked down as gut feeling and left at that.

The endpoint values represent the result of discounted cash flow analysis, discounted at the corporate cost of capital. The margin between the utility curve and the EMV curve is sometimes viewed as the *minimum risk premium*. This is one reason why negotiations get sticky on the subject of profitability. For drilling ventures that are successful, the rewards are spectacular. But only within the narrow context of the discovery well which requires ignoring any associated dry holes. Once a discovery is made it is hard to speak in terms of chance of success. This is particularly true from a government's viewpoint because they may not be aware or may not care that the company may have drilled five dry holes prior to the discovery.

MONTE CARLO SIMULATION

Monte Carlo Simulation holds an interesting place in the firmament of risk analysis. In some respects, it captures the essence of quantifying risk. And, like many new tools, it has changed the language of risk analysis and has altered the landscape of oil and gas reserve definitions.

Deterministic vs. Stochastic (Probabilistic) Approach

In the language of risk analysis the essence of characterizing uncertainty is to preserve the uncertainty all the way through to the "answer." The first attempt to characterize the range of possible outcomes is the age-old minimum, most likely, and maximum estimates. However, with the deterministic approach the calculation of the minimum estimate is typically based upon the minimum value for each possible variable. This provides what might be considered absolute minimum and maximum values. These absolute extremes have little value from a management perspective because they are too far-out.

Mother Nature does not work that way. The likelihood that a possible outcome could result from every variable being at it's minimum or maximum value is remote at best. What is needed is a reasonable minimum and a reasonable maximum. This would provide management with much more useable tools. Simulation is one answer to this problem. The following tables and figures compare ranges of reserve estimates (Table 6–6, Figs. 6–14, 6–15, 6–16).

Deterministic Approach

	Recoverable Oil(MMBBLS)	
Minimum	Most Likely	Maximum
22	48	79

Probabilistic Approach

	Recoverable Oil (MMBBLS)	
1st Decile (P10)	50th Percentile (P50)	9th Decile (P90)
37	46	55

Table 6–6 Example Reserves Estimates (See Figure 6-16)

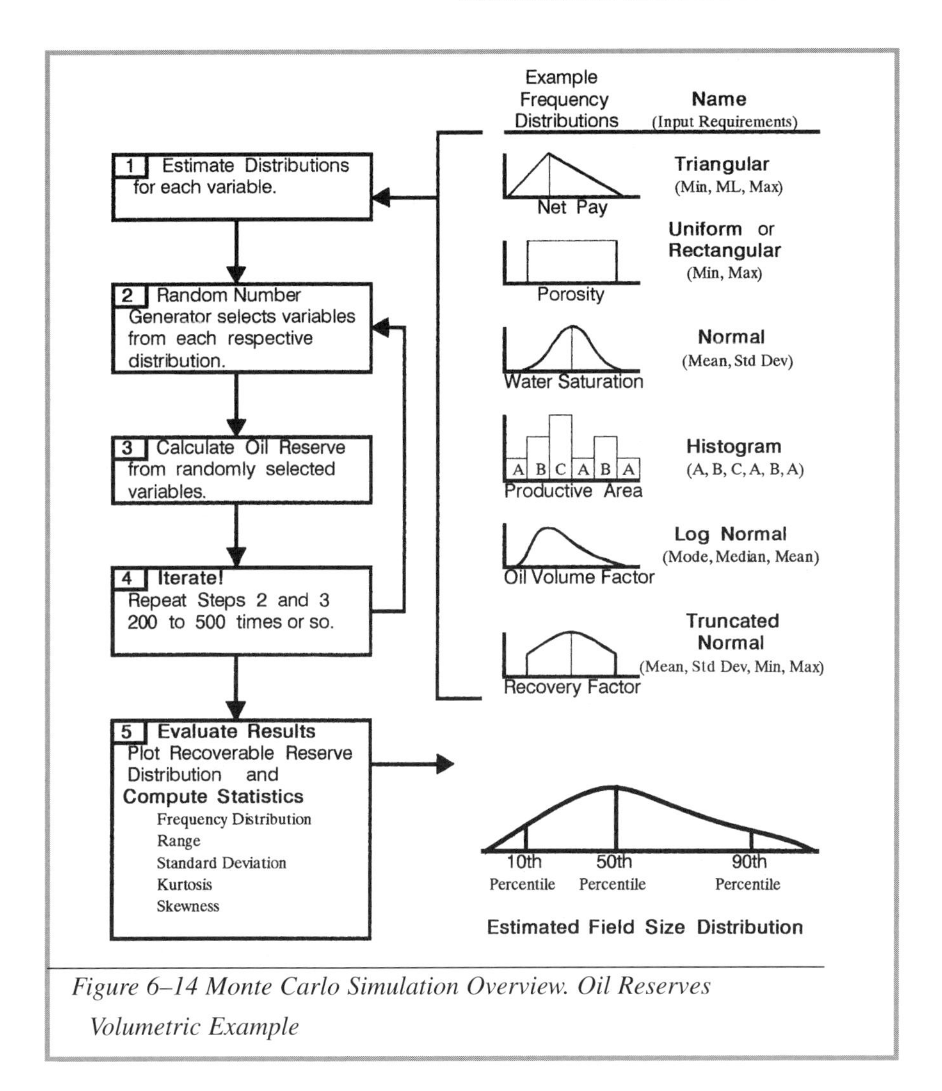

Figure 6–14 Monte Carlo Simulation Overview. Oil Reserves Volumetric Example

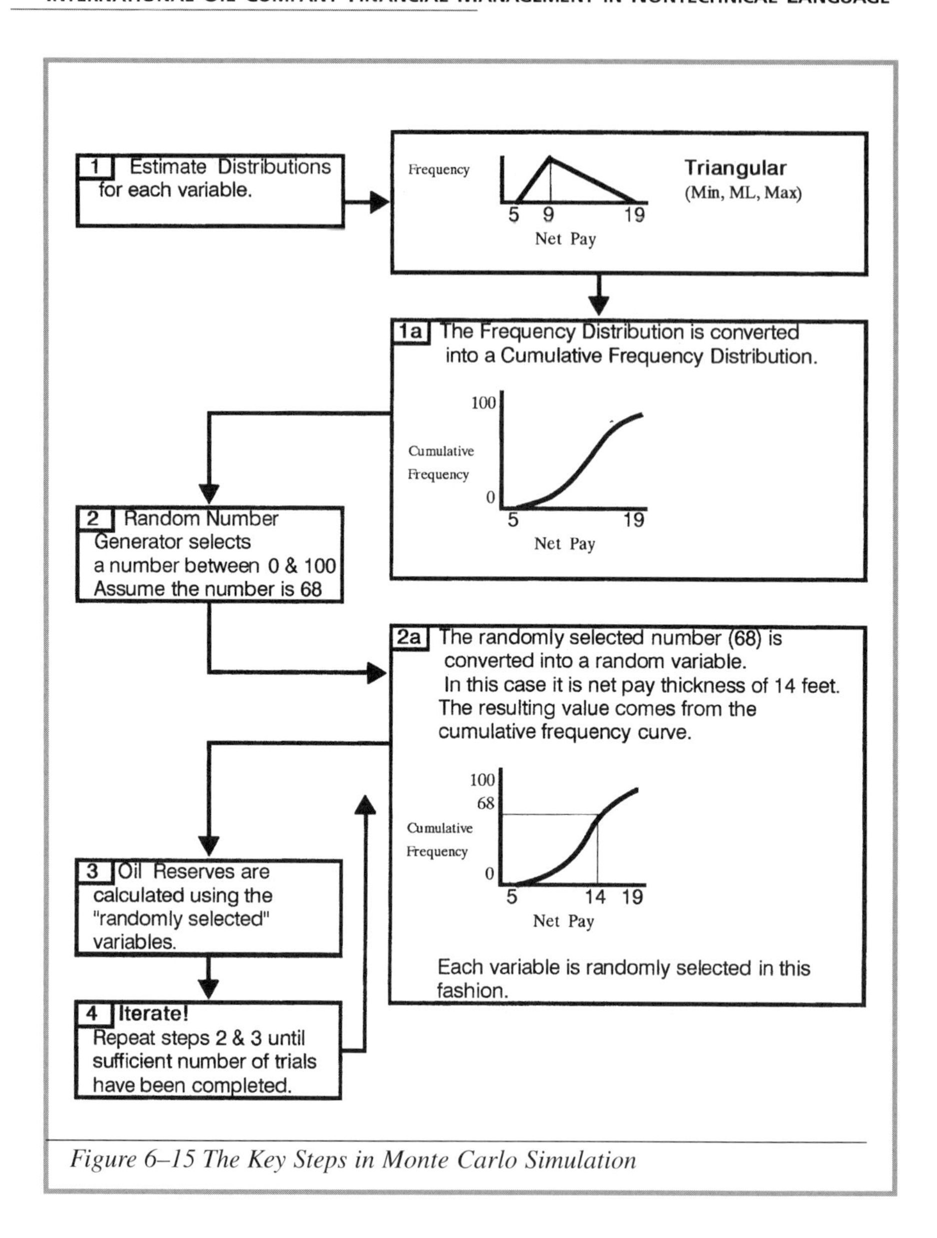

Figure 6–15 The Key Steps in Monte Carlo Simulation

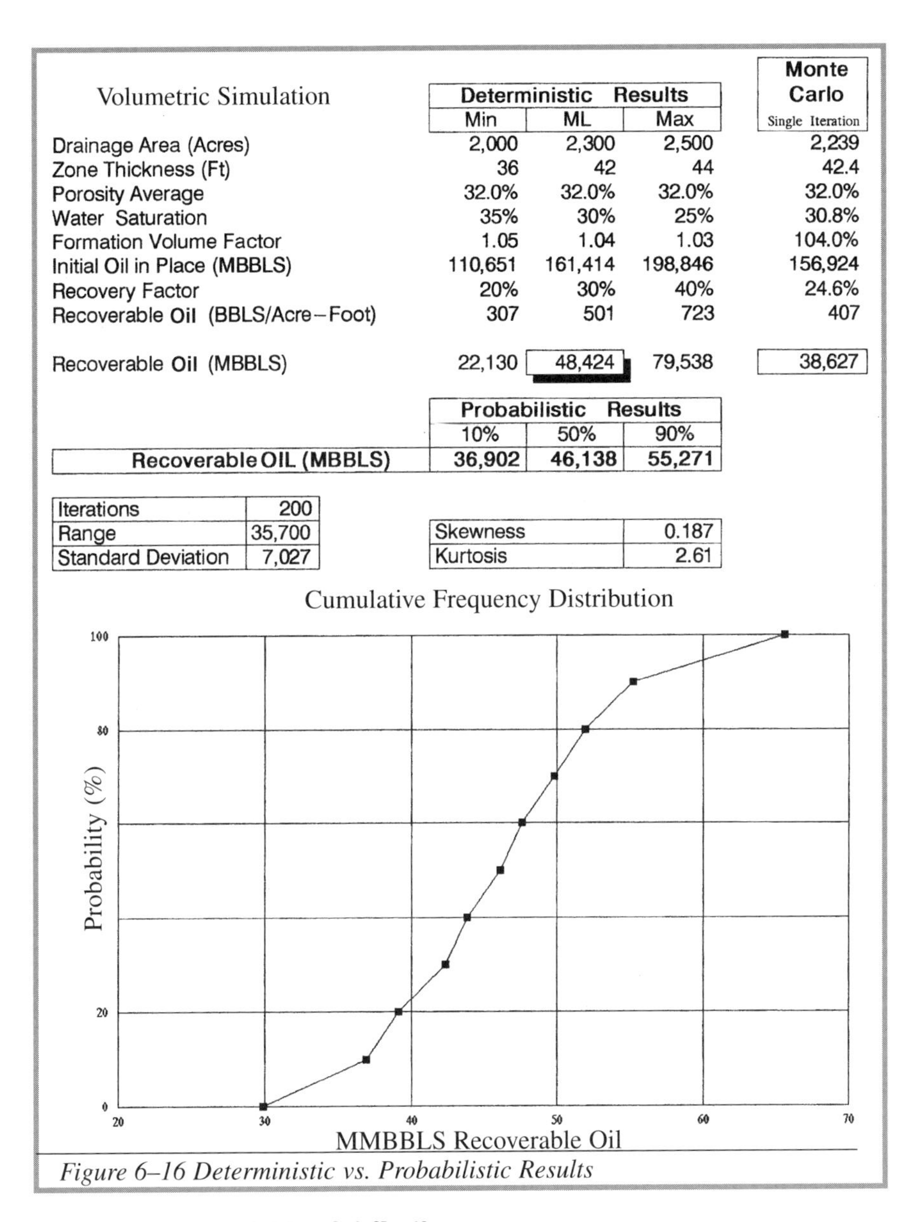

Volumetric Simulation

	Deterministic Results			Monte Carlo
	Min	ML	Max	Single Iteration
Drainage Area (Acres)	2,000	2,300	2,500	2,239
Zone Thickness (Ft)	36	42	44	42.4
Porosity Average	32.0%	32.0%	32.0%	32.0%
Water Saturation	35%	30%	25%	30.8%
Formation Volume Factor	1.05	1.04	1.03	104.0%
Initial Oil in Place (MBBLS)	110,651	161,414	198,846	156,924
Recovery Factor	20%	30%	40%	24.6%
Recoverable Oil (BBLS/Acre-Foot)	307	501	723	407
Recoverable Oil (MBBLS)	22,130	48,424	79,538	38,627

	Probabilistic Results		
	10%	50%	90%
Recoverable OIL (MBBLS)	**36,902**	**46,138**	**55,271**

Iterations	200
Range	35,700
Standard Deviation	7,027

Skewness	0.187
Kurtosis	2.61

Figure 6–16 Deterministic vs. Probabilistic Results

Characterizing Variable Distributions

The characterization of variables can be as complex as anything. It is important to remember that the object of the exercise is to try to "preserve the uncertainty" so that the simulation results will as accurately as possible characterize the dispersion of possible results. Suppose an estimate of porosity is being made and the estimate is going to be based upon historical data from a particular reservoir.

Example 1	*Historical Data*								
Porosity (%)	*12,*	*14,*	*14,*	*14,*	*14,*	*14,*	*18,*	*20,*	*20,*	*14*
Average		*16.4%*								

Example 2	*Historical Data*								
Porosity (%)	*12,*	*12,*	*12,*	*14,*	*14,*	*14,*	*14,*	*18,*	*20,*	*24*
Average		*15.4%*								

	Min	*ML*	*Max*
Using a triangular distribution, both of these would be characterized by:	*12%*	*14%*	*24%*

This example is used to illustrate the kind of things to look for. Both distributions would probably be more accurately characterized by using a histogram distribution. Most modern simulation packages will allow this type of distribution input (Figs. 6–17, 6–18).

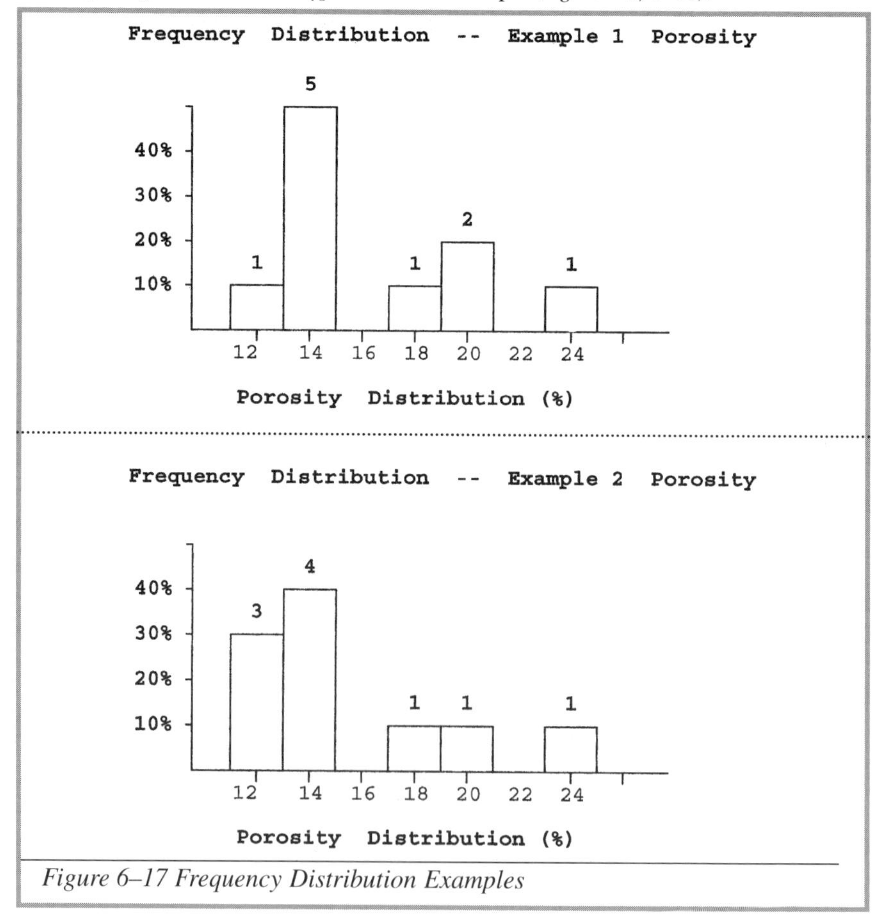

Figure 6–17 Frequency Distribution Examples

Variable Dependency Relationships

The example of a recoverable oil reserve estimate with a Monte Carlo Simulation (MCS) model provides a perfect example of a critical aspect of numerical simulation. There is a relationship between porosity and water saturation. Generally, the better the porosity, the lower the water saturation. In the language of MCS, there is a dependency relationship between these two variables. Water saturation is partially dependent on porosity. It is not an independent variable. Therefore, the model should account for this relationship.

Monte Carlo Arithmetic (1 + 1 = 3)

One drawback to oil and gas reserves analysis with Monte Carlo Simulation stems from the definition of proved reserves. The conventional approach these days is to assign the lowest 10th percentile (or first decile) as proved (P), the 50th percentile as proved + probable (P+P), and the upper 10th percentile (top decile) as proved + probable + possible (P+P+P). For example, in the Monte Carlo sample above, the various confidence intervals would correspond as follows:

10% Confidence	*36.9 million barrels*	=	*P*
Mean	*46.1 million barrels*	=	*P + P*
90% Confidence	*55.3 million barrels*	=	*P + P + P*

If two such fields were part of an oil company's portfolio of properties, then the reserves for the two fields should be:

10% Confidence	*73.8 million barrels*	=	*P*
Mean	*92.2 million barrels*	=	*P + P*
90% Confidence	*110.6 million barrels*	=	*P + P + P*

No, they would not!

It may be best to use an example to illustrate the problem. If we roll a six-sided die the probability of getting a value of 2 or more is 83% (5/6). We would be 83% confident that with each roll of the dice, the value would be equal to or greater than 2. This approach is similar to that of estimating proved reserves by using a 90% confidence level that there is a 90% chance that the reserves will be equal to or greater than the proved estimate. If we roll two dice each with an 83% chance of yielding a value of 2 or more, then is there an 83% chance that the combined value will be 4 or greater? No! The probability that the sum of two dice thrown being equal to 4 or more is closer to 92% (33/36). Much more conservative than 83%. This same thing happens with probabilistic reserves estimates. Adding probabilistic reserves estimates only works for the means or 50th percentile values.

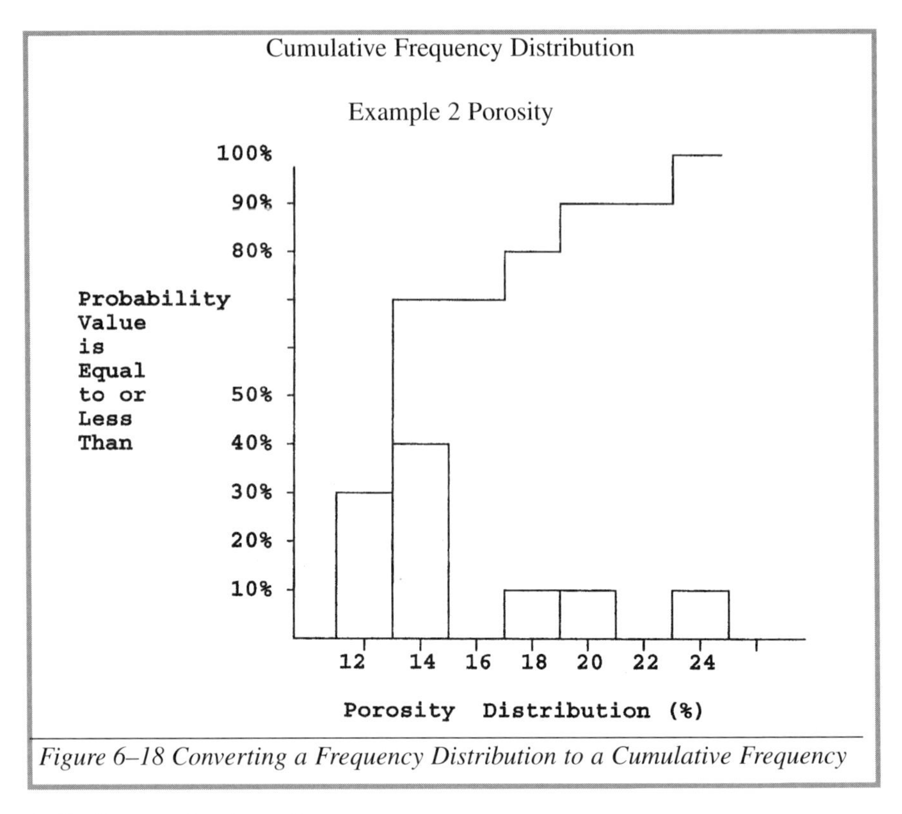

Figure 6–18 Converting a Frequency Distribution to a Cumulative Frequency

Latin Hypercube Sampling

Some programs offer an optional sampling technique that is more than what is ordinarily used and referred to as Monte Carlo sampling. The Latin Hypercube technique is the most common and is touted as having the ability to yield statistically significant samples and similar results with fewer iterations hence less computer time. The Latin Hypercube technique is a stratified sampling technique. It effectively forces the random number generating mechanism of the MCS program to select numbers from specific strata on the cumulative frequency distribution curves so that a sufficiently broad spectrum of random variables are obtained with the least effort.

PRESENTATION TO MANAGEMENT

The best analysis in the world will fail if it is not expressed clearly, both orally and in writing. The format of the report should accomplish two things: minimize decision maker's time in getting the meat out of the report, and presenting the important items of a rigorous evaluation report. The report should be organized so that one can read a limited amount of the report and understand the reason for the report, the important conclusions, and the recommendations without having to spend an inordinate amount of time. The report should have an introduction that tells the uninformed reader the reason and objective for the report. A numerical listing of the important conclusions is often preferred next. The conclusions should be written in order of importance with the most important being first. For an oil property evaluation, the conclusions probably would incorporate the following:

1. Present value of the future production discounted to the effective date of the study.
2. Estimated remaining reserve (ERR) and estimated ultimate recovery (EUR) at the effective date of the study.

Project Evaluation Checklist

Evaluation Date	________
Type of Evaluation	________
Effective Date of Analysis	________

	Min (P10)	ML (P50)	Max (P90)
Potential Recoverable Reserves			
Oil (MMBBLS)	____	____	____
Gas (BCF)	____	____	____
BOE (MMBBLS) (6:1)	____	____	____
Net Present Value (15%)	____	____	____
Internal Rate of Return	____	____	____
Payout (Years)	____	____	____
Return on Investment	____	____	____
Maximum Cash Impairment	____	____	____
Risk Capital ($MM)	____	____	____

	Most Likely Case	
	Project	$/Unit
Cumulative Net Cash Flow	____	____
Present Value (10% DCF)	____	____
(15% DCF)	____	____
(20% DCF)	____	____

Peak Production Rate (BOPD/MMCFD)	____
Liquid Yield (BBLS/MMCF)	____
GOR (Cubic Feet/BBL)	____
Capital Costs ($/BBL)	____
Government Take (%)	____

Success Ratios

Technical vs Commercial	________
Exploratory vs Development	________
Threshold field size Development	____ (MMBOE)
Threshold field size Exploration	____ (MMBOE)

Sensitivity Analysis

Product Prices	Ultimate Recovery
Capital Costs	Deliverability
Operating Costs	Timing Delays

Fig. 6–19 Presentation to Management

Recommendations and discussion of data should follow the conclusions. Fig. 6-19 provides an evaluation checklist of information that should be included in the presentation.

For further illustration, Figure 6-20 depicts the global market for exploration and development projects. Fiscal terms are discussed in Chapter 7.

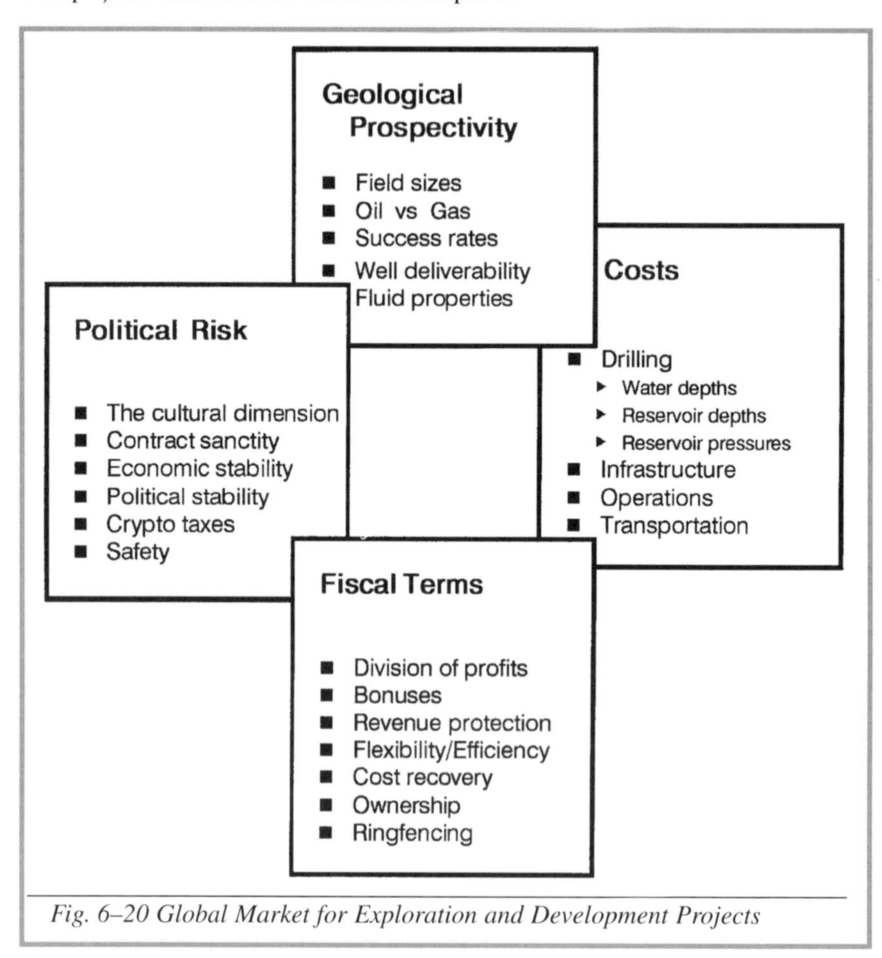

Fig. 6–20 Global Market for Exploration and Development Projects

Petroleum Fiscal Systems

7

The key differences in the international oil patch are *negotiations and fiscal terms, geopotential, and costs*. The bulk of this chapter deals with the commercial aspects of a given fiscal system. While engineering, economic, and geological principles are fairly universal, there are some substantial differences between the domestic U.S. and the international sector.

NEGOTIATIONS AND FISCAL TERMS

Many government and company negotiators try to appreciate the views of their counterparts. But it is not easy. There are four main issues that are the focus of concern for most governments or national oil companies:

- Keeping costs under control
- Maximum efficient production rates
- Evaluating the companies visiting their country
- The terms

While there are numerous concerns, the following are some of the most dramatic.

Cost Control

There appears to be a morbid fear in most national oil companies (or ministries) that oil companies will waste the natural resources through frivolous spending as well as cheating on costs, i.e., claiming they spent more than they spent. Yet, governments hold most (but not all) of the cards on this issue. Their primary means of protection include:

- The budget process
- Authorizations for expenditure (AFEs)
- Procurement processes and thresholds
- Audits (government auditors can be ruthless)
- Laws and penalties for noncompliance or fraud

- Natural desire and inclination of oil companies to economize due to profit sharing mechanisms in all systems.
- Nonoperator partners—watchdogs

Maximum Efficient Rate

The conventional wisdom in almost all governments (aside from the odd engineer in the national oil company) is that an oil company's only concern is short-term profits, and that they will produce oil too quickly and inefficiently, leaving behind more oil than they otherwise would. This is a very common point of debate and is often a very emotional issue for both sides.

Evaluating The Companies

The negotiators from the major and large independent oil companies do not have to deal with this issue so much. However, small independent oil companies from North America and elsewhere must spend a lot of effort demonstrating financial strength and technical ability. Government officials often have huge concerns about "unknown" entities and how to perform due diligence on these companies.

Fiscal Terms

Economic theory focuses on the produce of the earth derived from labor and capital. Rent theory deals with how this produce is divided among the laborers, owners of the capital and landowners through wages, profit, and rent.

In its most common usage in the petroleum industry, the term *economic rent* represents the difference between the value of production and the costs to extract it. These costs consist of normal exploration, development, and operating costs as well as an appropriate share of profit for the petroleum industry. Rent is the surplus. Economic rent is synonymous with excess profit. Governments attempt to capture as much economic rent as possible through various levies, taxes, royalties, and bonuses (Fig. 7–1).

The objective of any host government is to design a fiscal system where exploration and development rights are acquired by those companies who place the highest value on them. In an efficient market, competitive bidding can help achieve this objective. From the perspective of many governments, the market is much more efficient in the 1990s than it was in the 1980s. There are simply more oil companies seeking investment opportunities.

How governments extract economic rent is important. The industry is particularly sensitive to certain forms of rent extraction such as bonuses and royalties that are not based on profits. Royalties are of particular concern to the industry because the rate base for royalties is gross revenues. A spectrum of elements that make up rent is illustrated in Figure 7–1. The nonprof-

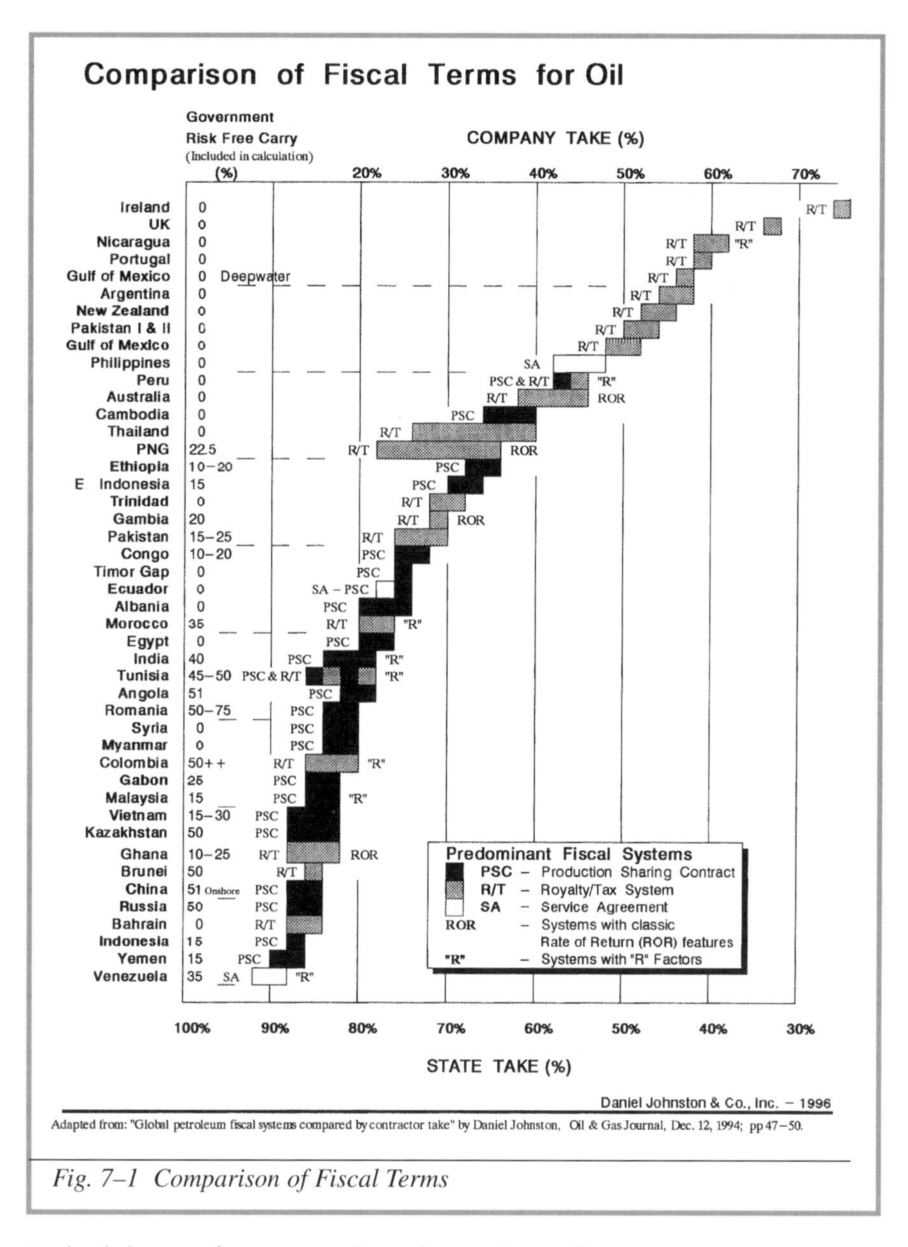

Fig. 7–1 Comparison of Fiscal Terms

it-related elements of government take, such as royalties and bonuses, are quite regressive—the lower the project profitability, the higher the effective taxes and levies. The further downstream from gross revenues a government levies taxes, the more progressive the system becomes. This is becoming more common. Royalties are being discarded in favor of higher taxes. This has advantages for both governments and the petroleum industry. However, there will always be governments that prefer some royalty or royalty equivalent. Royalties provide a guarantee that the government will benefit in the early stages of production.

There is the natural human fear of signing a bad deal and that the other party has negotiated a windfall for themselves. Government officials do their negotiating on a stage that is viewed by other government agencies, the ruler, dictator or prime minister, the citizens and the press. They are usually under a lot of pressure and have substantial political context within which they must make their decisions. The mechanisms that determine the division of profits are the focus of much of the negotiations. Beyond that, nearly every government wants to ensure for themselves a share of revenues in any given accounting period. Some governments refer to this as *revenue protection*—oil companies view these mechanisms as *royalty equivalents*. Royalties will certainly guarantee the government a share of revenues (or production) in any given accounting period, but royalties are not the only mechanisms that do this.

Analysis of the commercial terms of any fiscal system or contract requires detailed cash flow modeling and analysis as well as a bit of art. When the exercise can distill the analysis into the following elements, various contracts/systems can be efficiently compared. The key commercial aspects of a fiscal system are:

- Signature bonus, + [(1)]
- Contractor take / government take
- Government carry
- Royalty +
- Cost recovery limit +
- Access to gross revenues (AGR)/revenue protection (RP)
  (RP is the equivalent of an "effective royalty rate")
- Entitlement index
- Ringfencing
- Crypto taxes *(see examples)
- Government grief index (GGI)*

(1) While bonuses are captured in many "take" statistics, they never really get full credit in either the "take" statistics or the AGR/RP statistics.

\+ While some of these elements are embodied in the take and revenue protection statistics, they still provide additional insight.

* Qualifiable, not quantifiable, this stuff is nearly pure gut feel. The term GGI is borrowed from Richard Barry's book *The Management of International Oil Operations*. It is a useful and self-descriptive term.

The hard part is to combine these elements into a meaningful comparison with the geopotential, coupled with the cost of doing business and government grief index (political risk is part of this of course). It is a challenge, and the marketplace is more dynamic than ever.

RP or the "equivalent royalty rate," is the absolute minimum share of revenues the government will get in any given accounting period. AGR is the absolute maximum share of revenues an oil company can get in any given accounting period. For example, government revenue protection is 14% in the Indonesian standard oil contract, thus an oil company AGR is 86%. If an oil company had a 40% working interest, the maximum share of revenues in any given accounting period would be 34.4% (86% of 40%). The entitlement index represents what percentage of proved recoverable reserves might be "booked" according to standard Securities and Exchange Commission (SEC) criteria.

Contractor/Government Take—The Common Denominator

Division of profits boils down to what is called contractor and government take. These are expressed as percentages. Contractor take is the percentage of profits to which the contractor is entitled. Government take is the complement of that.

Contractor take provides an important comparison between one fiscal system and another. It focuses exclusively on the division of profits, and correlates directly with reserve values, field size thresholds and other measures of relative economics. Under a system such as the Indonesian production sharing contract with the well-known 85/15% split in favor of the government, the contractor may still end up with a 35% to 50% share of production. The Indonesian 15% contractor take is a measure of the contractor share of profits. It is a meaningful number even though most people know it is really closer to an 87/13 split. The formulas for calculating take are as follows:

Operating income ($)	=	*Cumulative gross revenues less cumulative gross costs over life of the project*
Government income ($)	=	*All government receipts from royalties, taxes, bonuses, production or profit sharing, etc.*
Government take (%)	=	*Government income divided by operating income*
Contractor take (%)	=	*(100% – government take)*

The best way to calculate take requires detailed economic modeling using cash flow analysis. Once a cash flow projection has been performed, the respective takes over the life of the project can be evaluated.

Under systems with liberal cost recovery provisions, the government take comes at a later stage of production than under those systems with royalties and restrictions on cost recovery. An even more detailed analysis of the respective takes would include the present value dimension. But this is not ordinarily done. Unfortunately, there is no standardization of terminology—numerous variations exist. Some of these are summarized in Table 7–11 at the end of this chapter.

A method is shown below for estimating government/contractor take and characterizing much of the financial elements of a system without detailed cash flow modeling. There are limitations to a quick-look approach such as this, but 95% of the time it provides extremely valuable information, and it is not difficult to do. The main limitation in estimating take is that it is not always easy to account for other aspects of a given fiscal system such as cost recovery limits, investment credits, royalty or tax holidays, domestic market obligations (DMOS), and crypto taxes.

Another dimension of contractor take comes from the effect of royalties or any taxes that are levied on gross revenues instead of profits. With different levels of profitability, fiscal systems with royalties can yield different government/contractor takes. This is shown in Table 7–1. This is a regressive fiscal structure. The lower the profitability, the higher the effective tax rate. This is because the royalty is based on gross revenues.

Case 1	Case 2	Case 3	
100.00	100.00	100.00	Gross Revenues
-10.00	-10.00	-10.00	**Royalty 10%**
	90.00	90.00	Net Revenues
	-30.00	-30.00	Costs
90.00	60.00	60.00	Taxable Income
-36.00	-24.00	-24.00	**Income Tax 40%**
54.00	36.00	36.00	Contractor Group Net Cash Flow
		-18.00	**Govt. Carry 50%**
		18.00	Contract or Net Cash Flow
	51.4%	25.7%	**Contractor Take**
54%			Contractor Marginal Take
46%			Govt. Marginal Take
	48.6%		Govt. Take
		74.3%	State Take

Table 7–1 Same Fiscal System—Different "Takes"

Marginal Take

The *net take on the marginal barrel* or *marginal take* approach is a slightly abstract view of the "take" concept which effectively ignores the cost element. This approach distorts somewhat the actual division of profits.

Marginal contractor take will almost always be equal to or greater than ordinary estimates or calculations of take. For those fiscal systems with no royalty, marginal take equals take. For example, in the UK where this terminology is used most often, there is no royalty. Contractor take is 67%. Marginal contractor take is also 67%.

Government vs. State Take

Government take effectively ignores the government risk-free carry through exploration. On the other hand, state take directly includes the government's share of profits resulting from the working interest obtained at the point of commerciality. There are numerous terms dealing with division of profits. Other terms are summarized in Table 7-11 at the end of this chapter.

About 40% of the countries in the world have the option of participating as a working interest partner at the point of discovery. This is ordinarily referred to as *government participation* or the *government carry*. For those countries that have this option the average carried interest is around 30%. The range is generally from 15 to 50%. At the point of discovery (usually plus a delineation well or two—the point of *commerciality*) the government takes up its working interest share and "pays its way" from that point forward. And, typically, the contractor is reimbursed for the government's share of past costs. It adds an interesting dynamic to fiscal system analysis and creates a bit of controversy. There are a number of analysts who believe that the government carry should be treated effectively as though it were a tax.

Access To Gross Revenues

An important perspective is the concept of access to gross revenues. Under a royalty/tax system with no formal cost recovery limit, companies are normally able to recover costs out of net revenue. Access to gross revenues is therefore limited only by the royalty. If there are sufficient eligible deductions, the company would pay no tax in a given accounting period. In most royalty/tax systems, in any given accounting period, there is no limit to the amount of deductions that a company can take and companies may be in a no-tax paying position.

Under a production sharing system with a 60% cost recovery limit, the contractor can recover costs of up to 60% of gross revenues in a given accounting period. Unrecovered costs are ordinarily carried forward. However, the contractor gets a share of the remaining 40% (profit oil). Therefore, the contractor's access to gross revenues will exceed the cost recovery limit.

Production Sharing Contract Structure

Take Calculation Full Cycle	Access to Revenues Calculation, Single Accounting Period	
100.0%	100.0%	Gross Revenues
-10.0	-10.0	Royalty
90.0	90.0	Net Revenues
-30.0	-60.0 *	Cost Recovery (60% Limit)
60.0	30.0	Profit Oil
-36.0	-18.0	60% to Government
24.0	12.0	40% to Contractor
-7.2	-0	Corp. Income Tax 30%
16.8	12.0	Contractor Net Income After Tax
28%		Contractor Take Contractor Net Income After Tax ÷ Gross Revenues – Costs
	72%	Access to Gross Revenues (Cost Oil + Profit Oil) (60 +12)
54%		Entitlement Index (Cost Oil + Profit Oil) (30 + 24)
	28%	Government Revenue Protection Effective Royalty Rate

*Costs must equal Full Cost recovery limit for AGR calculation.

Table 7–2 Access to Gross Revenues Calculation

Both royalties and cost recovery limits will guarantee that in any given accounting period the government will get a share of gross revenues regardless of whether true economic profits have been generated. At the end of the life of an oil field, the government guaranteed share of gross revenues takes on all of the characteristics of a royalty and many fiscal systems in this world need re-thinking in this respect.

The calculation of AGR requires the assumption that costs and eligible expenditures or deductions exceed gross revenues. Therefore, cost recovery is at the maximum and deductions for tax calculations exceed taxable income. Examples are shown in Tables 7-2 and 7-3.

Implications of Contractor Take

In 1993 an independent U.S. oil company made a 25 million barrel discovery off the Northwest Shelf of western Australia. The discovery was not exactly headline news outside western Australia. The fiscal terms in Australia are quite good, though—more than three times better than in Indonesia or Malaysia. Therefore, discovery amounted to the equivalent of an 80 million barrel Indonesian or Malaysian field. The fiscal terms make a huge difference.

The level of contractor take also has a direct impact on the value of reserves. Based on a wellhead price of $18.00 per barrel the value of provcd developed producing (PDP) working interest reserves in Indonesia is $1.10 to $1.60 per barrel. In the United States the value of similar reserves would be $4.50 to $6.00 per barrel.

Most of the science of fiscal system analysis deals with exploration economics and, to a lesser extent, development and production economics.

Country	Type of System	Royalty	Cost Recovery Limit	Access to Gross Revenues
UK	R/T	0%	100%	100%
NewZealand	R/T	5% (1)	100%	95%
Philippines	S/A	-7.5% (2)	70%	87%
Indonesia	PSC	0%	80%	86%
WorldAverage	N/A	7%	80%	81%
Malaysia	PSC	10.5% (3)	50%	64%
Egypt	PSC	0%	±40%	±52%

R/T = Royalty Tax System
S/A = Service Agreement
PSC = Production Sharing Contract
N/A = Not Applicable

(1) 5% Ad Valorem Royalty (AVR)
(2) Filipino Participation Incentive Allowance (FPIA)
(3) 10% Royalty + 0.5% Research Cess (based on gross revenues)

Table 7–3 Access to Gross Revenues Examples

NEGOTIATIONS

Governments have devised numerous frameworks for extracting economic rent from the petroleum sector. Some are well-balanced, efficient, and cleverly designed. Some will not work. The fundamental issue is whether or not exploration and/or development is feasible under the conditions outlined in the fiscal system. The result of government efforts are sometimes referred to as fiscal marksmanship. Structuring a fiscal system that will be appropriate or on-target under a variety of unknown future circumstances is difficult.

The purpose of fiscal structuring and taxation is to capture all economic rent. This is consistent with giving the industry a reasonable share of profit or take. But the level of industry profit considered to be fair and reasonable is debatable. The issue of the division of profits lies at the heart of contract/license negotiations.

Government Objectives

The objective of a host government is to maximize wealth from its natural resources by encouraging appropriate levels of exploration and development activity. In order to accomplish this, governments must design fiscal systems that:

1. Provide potential for a fair return to the state and industry
2. Avoid undue speculation
3. Limit administrative burden
4. Provide flexibility
5. Create healthy competition and promote market efficiency

The design of an efficient fiscal system must take into consideration the political and geological risks as well as the potential rewards. Malaysia has one of the toughest fiscal systems in the world. But, Malaysia has good geological potential. Many companies would love to explore in Malaysia and the government knows this. Governments are not the only ones who draw the line between fair return and rent. The market works both ways.

One country may tax profits at a rate of 85% or more, like Indonesia, while another country may only have an effective tax rate of 40%, like Spain. Yet both countries may be efficiently extracting their resource rent regardless of the kind of system that is used.

Oil Company Objective

The objective of oil companies is to build equity and maximize wealth by finding and producing oil and gas reserves at the lowest possible cost and highest possible profit margin. In order to do this they must explore for huge fields. Unfortunately, the regions where huge fields are likely to be found are often accompanied by tight fiscal terms. The oil industry is comfortable with tough terms if they are justified by sufficient geological potential. This is the birthplace of dynamic negotiations.

The primary economic aspects of contract/license negotiations are the work commitment and the fiscal terms. These are sometimes collectively referred to as the *commercial terms*. The work commitment represents hard "risk dollars" while fiscal terms govern the allocation of revenues that may result if exploration efforts are successful (Fig. 7-2).

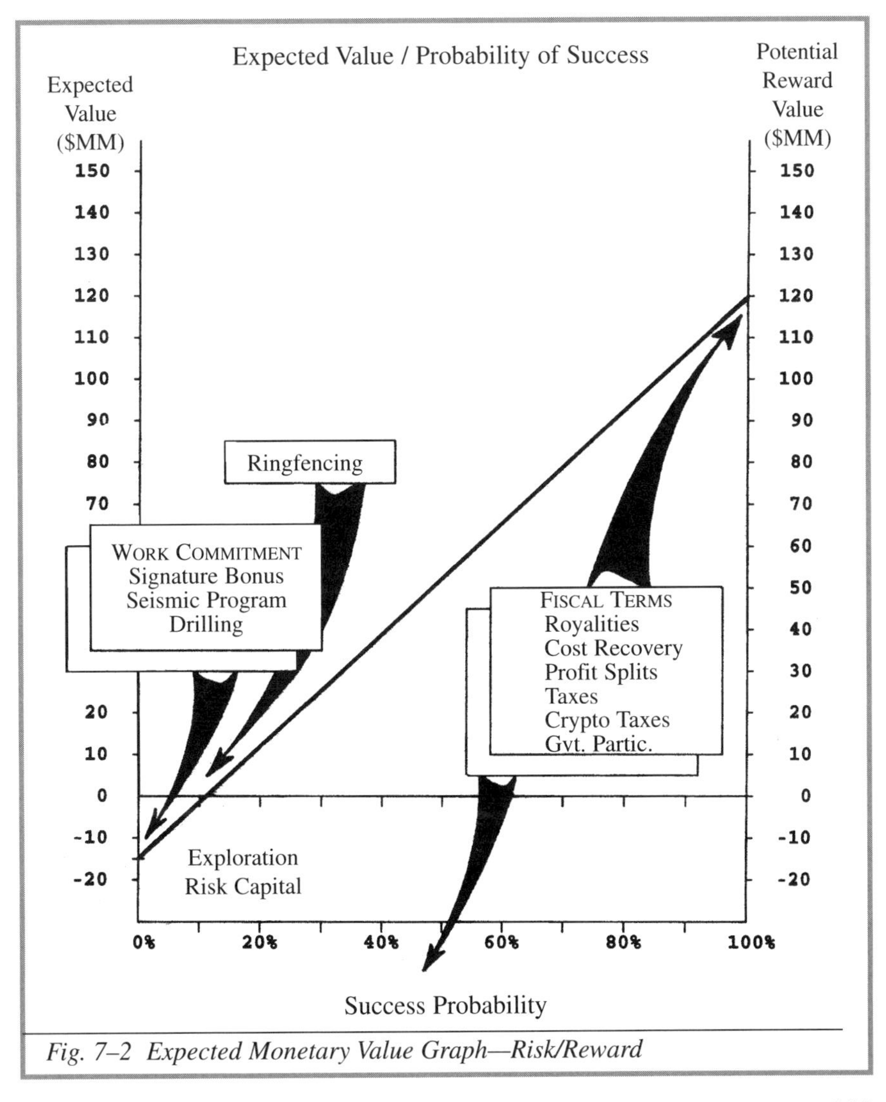

Fig. 7–2 Expected Monetary Value Graph—Risk/Reward

FAMILIES OF SYSTEMS

Governments and companies negotiate their interests in one of two basic systems: *concessionary* and *contractual*. The fundamental difference between them stems from different attitudes towards the ownership of mineral resources. The classification of petroleum fiscal systems is outlined in Figure 7-3.

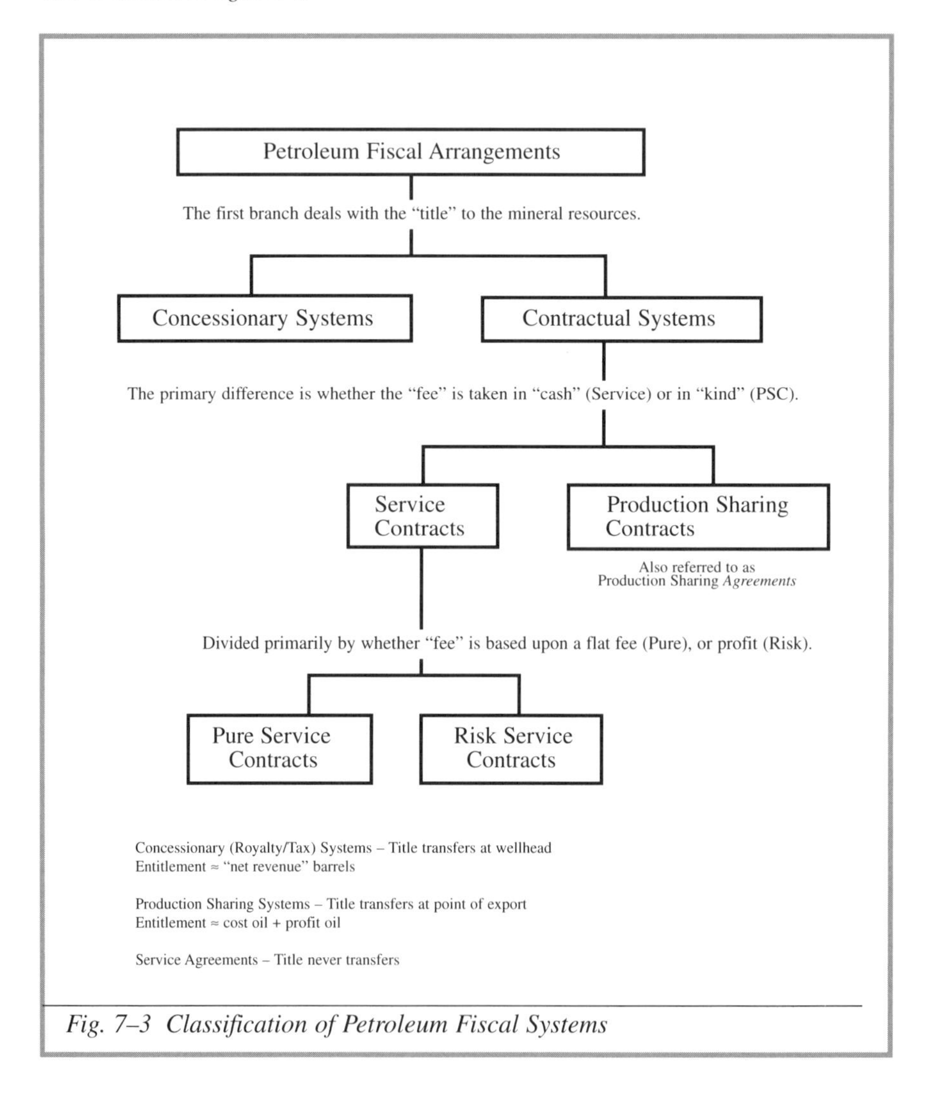

Fig. 7–3 Classification of Petroleum Fiscal Systems

Concessionary Systems

Concessionary systems, as the term implies, allow private ownership of mineral resources. Of course, the U.S. is the extreme example of such a system where individuals may own mineral rights. In most countries the government owns all mineral resources, but under concessionary systems it will transfer title of the minerals to a company if they are produced. The company is then subject to payment of royalties and taxes.

Contractual Systems

Under contractual systems the government retains ownership of minerals. Oil companies have the right to receive a share of production or revenues from the sale of oil and gas in accordance with a production sharing contract (PSC) or a service contract. In the petroleum industry, Indonesia is the pioneer of the PSC with the first contracts signed in the early and mid 1960s.

It is partially because of the concept of ownership of mineral resources that the term contractor has come into such wide use. The earliest uses of the production sharing concept occurred in the agriculture industry. Therefore, the term is used in the same context as sharecropper where ownership of the land and minerals is held by the government/landlord. The contractor or tenant/sharecropper is compensated out of production of minerals or grain according to a specific sharing arrangement. The term contractor therefore applies to PSCs or service agreements only, but with practical usage it cuts across the boundaries between PSCs and concessionary systems.

There is another aspect to this ownership issue. In most contractual systems, the facilities emplaced by the contractor within the host government domain become the property of the state either the moment they are landed in the country or upon start-up or commissioning. Sometimes title to the assets or facilities does not pass to the government until the attendant costs have been recovered. This transfer of title on assets, facilities and equipment does not apply to leased equipment or to equipment brought in by service companies.

Contractual arrangements are divided into service contracts and production sharing contracts. The difference between them depends on whether or not the contractor receives compensation in cash or in kind (crude). This is a rather modest distinction and as a result, systems on both branches are commonly referred to as PSCs, or sometimes production sharing agreements (PSAs). For example, in the Philippines the government alternately refers to their contractual arrangement as either a service contract or a PSC. The oil community does the same thing, but more ordinarily refers to this system as a PSC. In a strict sense though, this system is a risk service contract.

The transfer of title is effectively shifted from the wellhead under a concessionary system to the point of export under a PSC. With a service or risk service agreement, the issue of ownership is removed altogether. In a service contract the contractor gets a share of revenues, not production. With that in mind, the term revenue sharing or profit sharing contract would be appropriate, but these terms are not used. Despite the differences between systems, it is possible to obtain the same economic results under a variety of systems. As far as the calculations are concerned, there are so few material differences between one family and another that it is difficult to make generalizations about any given family of systems. Pure service agreements, rare as they are, are similar to arrangements found in the oil service industry.

Variation on Two Themes

Numerous variations are found in both the concessionary and contractual systems. The philosophical differences between the two systems have fostered a terminology unique to each. However, the terms are simply different names for basic concepts common to both systems. Because of the modest difference between service agreements and PSCs, the study of PSCs effectively covers the whole contractual branch of the tree. Much of the language and arithmetic of PSCs and service agreements is identical. The most dramatic differences between one fiscal system and another have to do with just how much taxation or government take is imposed.

CONCESSIONARY SYSTEMS

Concessionary arrangements predominated through the early 1960s. The earliest agreements consisted of only a royalty payment to the state. The simple royalty arrangements were followed by larger royalties. Taxes were added once governments gained more bargaining power. In the late 1970s and early 1980s, a number of governments created additional taxes to capture excess profits from unexpectedly high oil prices. Now there are numerous fiscal devices, layers of taxation, and sophisticated formulas found in concessionary systems.

Flow Diagram

Figure 7–4 depicts revenue distribution under a simple concessionary system. This example is provided to develop further the concept of contractor take and to compare with other systems. The diagram illustrates the hierarchy of royalties, deductions, and (in this example) two layers of taxation. For illustration purposes, a single barrel of oil is forced through the system.

Royalties

The royalty comes right off the top. In this example a 20% royalty is used. Gross revenue less royalty equals net revenue.

Deductions

Operating costs, depreciation, depletion, amortization (DD&A), and intangible costs (if applicable) are deducted from net revenue to arrive at taxable income. DD&A is the common terminology, but depletion is seldom allowed. When the term is used, it is assumed that it applies to depreciation and amortization. Most countries follow this format, but will have different allowed rates of depreciation or amortization for various costs. Some countries are liberal in allowing capital costs to be expensed and not forced through DD&A.

Taxation

Revenue remaining after royalty and deductions is called *taxable income* (Table 7–4). In Figure 7-4 it is subjected to two layers of taxation—10% provincial and 40% federal taxes. Provincial taxes are deductible against federal taxes. The effective tax rate therefore is 46%. With

Gross Revenues	=	Total Oil and Gas Revenues
Net Revenues	=	Gross Revenues
	–	Royalties
Net Revenue (%)	=	100%, Royalty Rate (%)
Taxable Income	=	Gross Revenues
	–	Royalties
	–	Operating Costs
Deductions	–	Intangible Capital Costs *
	–	DD&A
	–	Investment Credits (if allowed)
	–	Interest on Financing (if allowed)
	–	Tax Loss Carry Forward
	–	Abandonment Cost Provision
	–	Bonuses **
Net Cash Flow	=	Gross Revenues
(after-tax)	–	Royalties
	–	Tangible Capital Costs
	–	Intangible Capital Costs *
	–	Operating Costs
	–	Bonuses
	–	Taxes

* In many systems no distinction is made between operating costs and intangible capital costs and both are expensed.

** Bonuses are not always deductible.

Table 7–4 Basic Equations—Royalty/Tax Systems

tax deductions the contractor share of gross revenues is 64%. The profit in this example is $11.00, ($20.00 less $9.00 in costs). The contractor share of profits is $3.78. Contractor take therefore is 34.36%. This is different than profit margin which in this example is 18.9% ($3.78/$20.00).

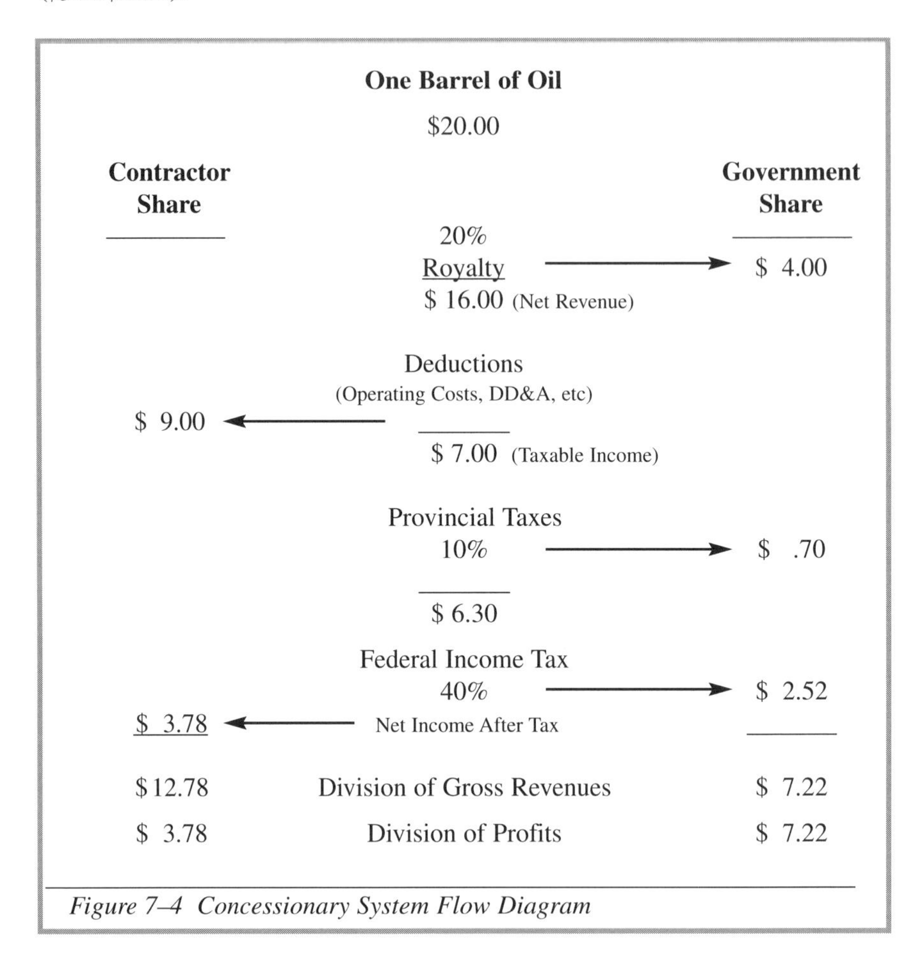

Figure 7–4 Concessionary System Flow Diagram

PRODUCTION SHARING CONTRACTS

At first PSCs and concessionary systems appear to be quite different. They have major symbolic and philosophical differences, but these serve more of a political function than anything else. The terminology is certainly distinct, but these systems are really not that different from a financial point of view. There is a general view that PSCs are more complex and onerous than concessionary systems. This is not a fair generalization. Too much diversity exists on both sides. The arithmetic of a simple PSC is evaluated first. Many of the other features of a

PSC are similar to those found under other systems. Therefore, these common elements are discussed in detail later. Of the numerous production sharing arrangements, there are common elements (Fig. 7-5). The essential characteristic, of course, is that of state ownership of

Gross revenues	=	Total Oil and Gas Revenues
Net revenues	=	Gross Revenues – Royalties
Net Revenue (%)	=	100% – Royalty Rate (%)
Cost Recovery	=	Operating Costs
"Cost Oil"	+	Intangible Capital Costs*
	+	DD&A
	+	Investment Credits (if allowed)
	+	Interest on Financing (if allowed
	+	Unrecovered Costs Carried Forward
	+	Abandonment Cost Provision
Profit Oil	=	Net Revenue – Cost Recovery
Contractor Profit Oil	=	Profit Oil x Contractor Percentage Share
Government Profit Oil	=	Profit Oil x Government Percentage Share
Net Cash Flow	=	Gross Revenues
	–	Royalties
	–	Tangible Capital Costs
	–	Intangible Capital Costs*
	–	Operating Costs
	+	Investment Credits
	–	Bonuses
	–	Government Profit Oil
	–	Taxes
Taxable Income	=	Gross Revenues
	–	Royalties
	–	Intangible Capital Costs*
	–	Operating Costs
	+	Investment Credits
	–	Government Profit Oil
	–	Abandonment Cost
	–	DD&A
	–	Bonuses**

* In many systems no distinction is made between operating costs and intangible capital costs and both are expensed.

** Bonuses are not always deductible.

Figure 7–5 Contractual System Basic Equation

the resources. The contractor receives a share of production for services performed. That share normally consists of two components—cost oil and profit oil (reimbursement and remuneration).

Production Sharing Contract Flow

Figure 7–6 shows a flow diagram of a PSC with a royalty and cost recovery limit. It illustrates the terminology and arithmetic hierarchy of a typical PSC. For illustration, one barrel of oil is used.

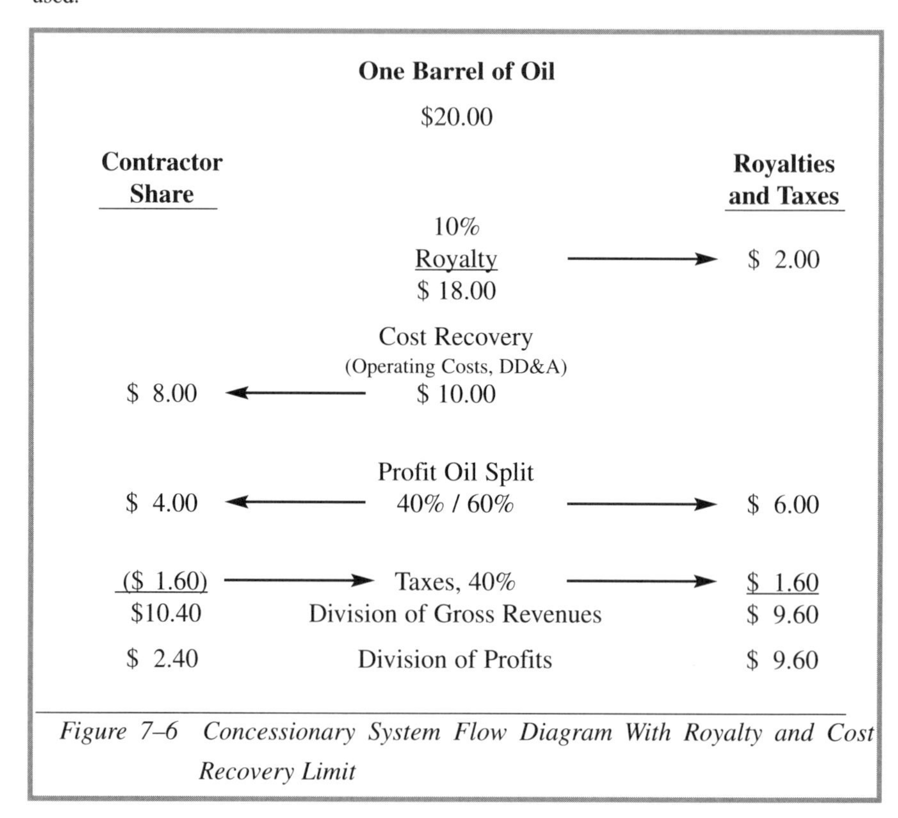

Figure 7–6 Concessionary System Flow Diagram With Royalty and Cost Recovery Limit

Royalty

The royalty comes right off the top just as it would in a concessionary system. The example in Figure 7–6 uses a 10% royalty.

Cost Recovery

Before sharing of production the contractor is allowed to recover costs out of net revenues. However, most PSCs will place a limit on cost recovery usually a percentage of gross revenues. For example, in Figure 7-6, cost recovery is limited to 40% of gross revenues. If operating costs and DD&A amounted to more than that, the balance would be carried forward and recovered later. From a mechanical point of view, the cost recovery limit is the only true distinction between concessionary systems and PSCs.

Profit Oil Split

Revenues remaining after royalty and cost recovery are referred to as *profit oil* or *profit gas*. The analog in a concessionary system would be taxable income. The terminology is precise because of the ownership issue. The term *taxable income* implies ownership that does not exist yet under a PSC. The contractor has nothing to tax. In this example, the contractor's share of profit oil is 40%.

Taxes

The tax rate is 40%. In the long run, for all practical purposes, the tax base is equal to the contractor profit oil.

Contractor Take

With cost recovery, the contractor's gross share of production comes to 52%. Total profit is $12.00. Considering the 10% royalty, profit oil split, and taxation, the contractor share of profits is $2.40. Contractor's take therefore is 20%.

BASIC ELEMENTS

The basic elements of a production sharing system are categorized in Table 7–5. These elements are also found in concessionary systems with the exception of the cost recovery limit and production sharing. Each of the economic elements listed in the table are discussed separately.

As Table 7-5 shows, many aspects of the government/contractor relationship may be negotiated, but some are normally determined by legislation. Those elements that are not legislated must be negotiated. Usually, the more aspects that are subject to negotiation, the better. This is true for the government agency responsible for negotiations as well as for the oil companies. Flexibility is required to offset differences between basins, regions, and license areas within a country.

Legislative bodies ordinarily have more authority than the national oil company or oil ministry empowered to negotiate contracts. Therefore, some items are simply not subject to negotiation. If a national oil company has authority to negotiate the profit oil split, then the tax rate

	National Legislation	Contract Negotiation
Revenue or Production Sharing Elements	Royalties* Taxation* Depreciation Rates Investment Credits Domestic Obligation Ringfencing	Bonus Payments* Cost Recovery Limits Production Sharing*
Other Aspects	Government Participation Ownership Transfer Arbitration Insurance	Work Commitment Relinquishment Commerciality

*Those features most commonly associated with contractor take.

Table 7–5 Typical Fiscal System Structure

is not such an issue. While fiscal elements such as taxes are normally legislated, others are subject to negotiation and are defined in the PSC. Only a contract can set forth such elements as contract area coordinates, work commitment, and duration of phases of the contract.

Work Commitment

Work commitments are generally measured in kilometers of seismic data and number of wells. There are some instances where the work commitment may consist only of seismic data acquisition with an option to drill. These are referred to as seismic options. Other contracts have hard, aggressive drilling obligations. The terms of the work commitment outline penalties for nonperformance. The work commitment is a critical aspect of international exploration. It embodies most of the risk of petroleum exploration.

Bonus Payments

Cash bonuses are sometimes paid upon finalization of negotiations and contract signing, hence the term signature bonus. Although cash payments are most common, the bonus may consist of equipment or technology. Not all PSCs have bonus requirements. Among contracts that have bonus provisions, there are many variations such as production bonuses, start-up bonuses, and cumulative production bonuses.

Royalties

Royalties are a fundamental concept, and the treatment is similar under almost all fiscal systems. While there are some exotic variations on the royalty theme, they are rare. Royalties are taken right off the top of gross revenues. Some systems will allow a netback of transportation costs. This may occur when there is a difference between the point of valuation for royalty calculation purposes and the point of sale. Transportation costs from the point of valuation to the point of sale are deducted (netted back).

In the Philippines, given sufficient level of Filipino ownership (30% onshore or 15% in deepwater), the government pays the contractor group 7.5% of gross revenues right off the top. This part of the contractor fee is equivalent to or may be viewed as a negative royalty.

A specific rate royalty is a fixed amount charged per barrel or per ton. These are relatively rare, but it may also go by another name, such *cess* or *export* tariff, like that found in the former Soviet Union (FSU). Another type of specific rate royalty is the $1.00/bbl (900 peso) War Tax levy in Colombia. These are even more regressive than an ordinary royalty.

Sliding Scales

A feature found in many petroleum fiscal systems is the sliding scale used for royalties, profit oil splits, and various other items. The most common approach is an incremental sliding scale based on average daily production. The following example shows a sliding scale royalty that steps up from 5% to 15% on 10,000 barrels of oil per day (BOPD) tranches of production. If average daily production is 15,000 BOPD, the aggregate effective royalty paid by the contractor is 6.667% (10,000 BOPD at 5% + 5,000 BOPD at 10%).

Sliding Scale Royalty Example

	Average Daily Production	*Royalty*
First Tranche	*Up to 10,000 BOPD*	*5%*
Second Tranche	*10,001 to 20,000 BOPD*	*10%*
Third Tranche	*Above 20,000 BOPD*	*15%*

Sometimes misconceptions arise when it is assumed that once production exceeds a particular threshold all production is subject to the higher royalty rate. Sliding scales do not work that way unless specified in the contract, and that is very rare.

Production levels in sliding scale systems must be chosen carefully. If rates are too high, then the system effectively does not have a flexible sliding scale. In some situations tranches of 50,000 BOPD can be too high, or conversely 10,000 BOPD tranches may be too low. The choice is dictated by the anticipated size of discoveries. For a point of reference, most fields worth developing in the international arena will produce from 15%-20% of their reserves in a peak year of production. Generally, the larger the field, the lower the peak percentage rate.

Therefore, a 100 million barrel field might be expected to produce perhaps around 15% of its reserves or 15 million barrels in the peak year of production. This is an average daily rate of 41,000 barrels oil per day. If a region is not capable of yielding greater than 100 million barrels fields, then sliding scale tranches of 50,000 barrels oil per day are useless.

The sequence of calculations that follow the royalty calculation always leads to the recovery of costs. Under the concessionary system, these are called *deductions*. Contractual systems use a more descriptive term called *cost recovery*.

Cost Recovery

Cost recovery is the means by which the contractor recoups costs of exploration, development, and operations out of gross revenues. Most production sharing contracts (PSCs) have a limit to the amount of revenues the contractor may claim for cost recovery but will allow unrecovered costs to be carried forward and recovered in succeeding years. Cost recovery limits or cost recovery *ceilings*, as they are also known, if they exist, typically range from 50%-70%.

Cost recovery is an ancient concept. Even communists are comfortable with it. The ones who put up the capital should at least get their investment back. Beyond that, there is wide disagreement. The cost recovery mechanism is one of the most common features of a PSC. It is only slightly different than the cost recovery (deduction) mechanism used in most concessionary systems. Once the original exploration and development costs are recovered, operating costs comprise the majority of recovered costs. At this stage, cost recovery may range from 15% to 30% of revenues.

In most respects, cost recovery is similar to deductions in calculating taxable income under a concessionary system. The profit oil share taken by the government could be viewed as a first layer of taxation. However, the terminology is specific and refers back to the ownership issue. A contractor under a PSC does not own the production, and therefore, at the point of cost recovery has no taxable revenues against which to apply deductions. The government reimburses the contractor for costs through the cost recovery mechanism and then shares a portion of the remaining production or revenues with the contractor. It is only at that point then that taxation becomes an issue.

Tangible vs. Intangible Capital Costs

Sometimes a distinction is made between depreciation of fixed capital assets and amortization of intangible capital costs. Under some concession agreements, intangible exploration and development costs are not amortized. They are expensed in the year they are incurred and treated as ordinary operating expenses. Those rare cases where intangible capital costs are written off immediately can be an important financial incentive. Most systems will force intangible costs to be amortized. Therefore, recovery of these costs takes longer, with more revenues subject to taxation in the early stages of production.

Interest Cost Recovery

Sometimes interest expense is allowed as a deduction in about 20% of the systems in the world, both PSCs and concessionary systems. This can include interest during construction or a rate based on cumulative unrecovered capital. Under a PSC, this is referred to as interest cost recovery. Some systems limit the amount of interest expense by using a theoretical capitalization structure such as a maximum 70% debt. Some systems will limit the interest rate itself, regardless of the actual rate of interest incurred. China allows development costs at 9% deemed interest per year compounded annually and recoverable through cost oil.

General & Administrative Costs (G&A)

Many systems allow the contractor to recover some home office administrative and overhead expenses. Nonoperators are normally not allowed to recover such costs. Contractors in Indonesia are limited to 2% of gross revenues for G&A cost recovery. The 1989 Myanmar model contract had a sliding scale allowance for G&A based on total petroleum costs each year:

- The first $5 million U.S. dollars @ 4%
- The next $3 million U.S. dollars @ 2 %
- The next $4 million U.S. dollars @ 1%
- Over $12 million U.S. dollars @ 0.5%

In China, an annual overhead charge is allowed for offshore exploration at a rate of 5% on the first $5 million per year, dropping down to 1% for costs above $25 million. Development overhead charges are allowed at a rate of 2.5% on the first $5 million, dropping down to 0.25% on costs exceeding $25 million. Overhead for operating costs is allowed at a flat rate of 1.8%.

Unrecovered Costs Carried Forward

Most unrecovered costs are carried forward and are available for recovery in subsequent periods. The same is true of unused deductions. The term sunk cost is applied to past costs that have not been recovered. There are four classes of sunk cost:

- Tax loss carry forward (TLCF)
- Unrecovered depreciation balance
- Unrecovered amortization balance
- Cost recovery carry forward

These items are typically held in abeyance prior to the beginning of production. Many PSCs do not allow preproduction costs to begin depreciation or amortization prior to the beginning of production, so there is no TLCF. Bonus payments, though, may create a TLCF.

Exploration sunk costs can have a significant impact on field development economics and they can strongly affect the development decision. The importance of sunk costs and development feasibility centers on an important concept called *commerciality*.

The financial impact of a sunk cost position on the development decision can be easily determined with discounted cash flow analysis. The field development cash flow projection should be run once with sunk costs and once without. The difference in present value between the two cash flow projections is the present value of the sunk cost position. For example, if a company has $20 million in sunk costs and is contemplating the development of a discovery, the present value of the sunk costs could be on the order of $10 to $15 million. This would depend on whether or not the costs could be expensed or if they must be capitalized. It also depends on restrictions on cost recovery—primarily the cost recovery limit.

Abandonment Costs

The issue of ownership adds an interesting flavor to the concept of abandonment liability. Under most PSCs, the contractor cedes ownership rights to the government for equipment, platforms, pipelines, and facilities upon commissioning or startup. The government as owner is theoretically responsible for the cost of abandonment. Ideally, abandonment costs are recovered through cost recovery just as the costs of exploration, development, and operations are. Anticipated cost of abandonment is accumulated through a sinking fund that matures at the time of abandonment. The costs are recovered prior to abandonment so that funds are available when needed.

Profit Oil Split and Taxation

Profit oil is split between the contractor and the government, according to the terms of the PSC. Sometimes it is negotiable. Published government or contractor take figures refer to the after-tax split.

Explorers focus on geopotential and how it balances with fiscal terms and the cost of doing business. When evaluating fiscal terms, the focus is on division of profits—the government/contractor take. Geopotential, costs, infrastructure, political stability, and other key factors that influence business decisions are weighed against contractor take. The split in most countries ranges from just under 15% to over 55% for the contractor (Figure 7–1). Beyond these extremes are the exceptions that are becoming more and more rare.

Commerciality

An important aspect of international exploration is the issue of commerciality. It deals with who determines whether or not a discovery is economically feasible and should be developed. It is a sensitive issue. There are often situations where accumulated exploration expenditures are so substantial that by the time a discovery is made, these sunk costs have a huge economic impact on the development decision. From the perspective of the contractor, sunk costs will flow through cost recovery upon development (or will be used as deductions), and they can represent considerable value. But they represent a liability, or a cost, as far as the government is concerned. If cost recovery is too great, then depending upon the contractual/fiscal structure, the government may end up with only a small percentage of the gross production. The grant of commercial status marks the end of the exploration phase and the beginning of the development phase of a contract.

Government Participation

Approximately 40% of the countries in the world have an option for the national oil company to participate in development projects. Under most government participation arrangements, the contractor bears the cost and risk of exploration, and if there is a discovery, the government backs-in for a percentage. In other words the government is carried through exploration. This is fairly common and automatically assumed whenever some percentage of government participation is quoted.

The financial effect of a government partner is similar to that of any working interest partner with a few large exceptions. First, the government is usually carried through the exploration phase and may or may not reimburse the contractor for past exploration costs, although they usually do out of their share of production. Second, the government's contribution to capital and operating costs is normally paid out of production. Finally, the government is seldom a silent partner.

Investment Credits And Uplifts

Some systems have incentives, such as investment credits or uplifts. Uplifts and investment credits are two names for the same basic concept. An uplift allows the contractor to recover an additional percentage of capital costs through cost recovery. It works the same way in a concessionary system. For example, an uplift of 20% on capital expenditures of $100 million would allow the contractor to recover $120 million.

Domestic Obligation

Many contracts have provisions that address the domestic crude oil or natural gas requirements of the host nation. These provisions are often referred to as the domestic supply requirement or domestic market obligation (DMO). Usually, they specify that a certain percentage of the contractor's profit oil be sold to the government. The sales price to the government is usually at a discount to world prices. The government may also pay for the domestic crude in local currency at a predetermined exchange rate. Revenues from sale of domestic crude are normally taxable.

Ringfencing

The issue of recovery or deductibility of costs is further defined by the revenue base from which costs can be deducted. Ordinarily, all costs associated with a given block or license must be recovered from revenues generated within that block. The block is *ringfenced*. This element of a system can have a huge impact on the recovery of costs of exploration and development. Indonesia requires each contract to be administered by a separate new company. This restricts consolidation and effectively erects a ringfence around each license area. Sixty percent of the countries are ringfenced.

Some countries will allow certain classes of costs associated with a given field or license to be recovered through revenues from another field or license. India allows exploration costs from one area to be recovered out of revenues from another, but development costs must be recovered from the license in which those costs were incurred. From the government perspective, any consolidation or allowance for costs to cross a ringfence means that the government may in effect subsidize unsuccessful operations. This is not a popular direction for governments because of the risky nature of exploration. However, allowing exploration costs to cross the fence can be a strong financial incentive for the industry.

If a country with an effective tax burden of 50% allowed exploration costs to be deducted across license boundaries, then the industry could be drilling with 50 cent dollars. It would cut the risk in half. From the perspective of the development engineer, it has little meaning unless development and operating costs are also allowed to cross. Dropping or loosening the ringfence can provide strong incentives, especially to companies that have existing production and are paying taxes.

Reinvestment Obligations

Some contracts require the contractor to set aside a specified percentage of income for further exploratory work within a license. In France, the level of taxation was effectively reduced when the company reinvested a certain portion of income. This approach is not as harsh as a firm obligation. The objective, of course, is to get companies to spend more in-country and repatriate less of their profits. Reinvestment obligations or reinvestment incentives are fairly rare.

Tax And Royalty Holidays

Governments can enact legislation or issue decrees that are designed to attract additional investment. Tax or royalty holidays are used for this purpose. These specify that for a given holiday period, royalty or taxes are not payable. For the 1989-90 licensing round, Myanmar built into their PSC a three-year holiday on their 30% income tax. The start of the holiday begins with the start of production. Therefore, the first three years of production from licenses issued at that time will not be burdened with corporate income tax.

When reviewing fiscal terms that have time limitations, the starting point is as important as the time period. In some instances the holiday begins on the effective date of the contract, or on a specific calendar date. In other cases the holiday may begin with production startup.

Geopotential

While geological and engineering principals are relatively universal, exploration potential is dramatically better outside most of the United States. This is due to the advanced maturity of the U.S. oil industry. Most foreign oil men are shocked to hear U.S. production statistics. For example, there are nearly 600,000 oil wells in the U.S. that produce an average of 11 barrels of oil per day. If the 1,500-odd Alaskan wells (around 1,000 barrels of

	Production Sharing Contracts	World Average System	Royalty Tax Systems
Number of Systems	68	123	55
Government Take **State Take**	66% 70%	62% 67%	58% 64%
Government Carry (40% have a carry)	12%	13%	15%
Royalty	5.7%	7.1%	8.9/%
Cost Recovery Limit	63%	N/A	N/A
Access to Gross Revenues	73%	81%	91%
Government Revenue Protection (Effective Royalty Rate)	27%	19%	10%
Entitlement Index	69%	79%	91%

Table 7–6 World Fiscal System Statistics

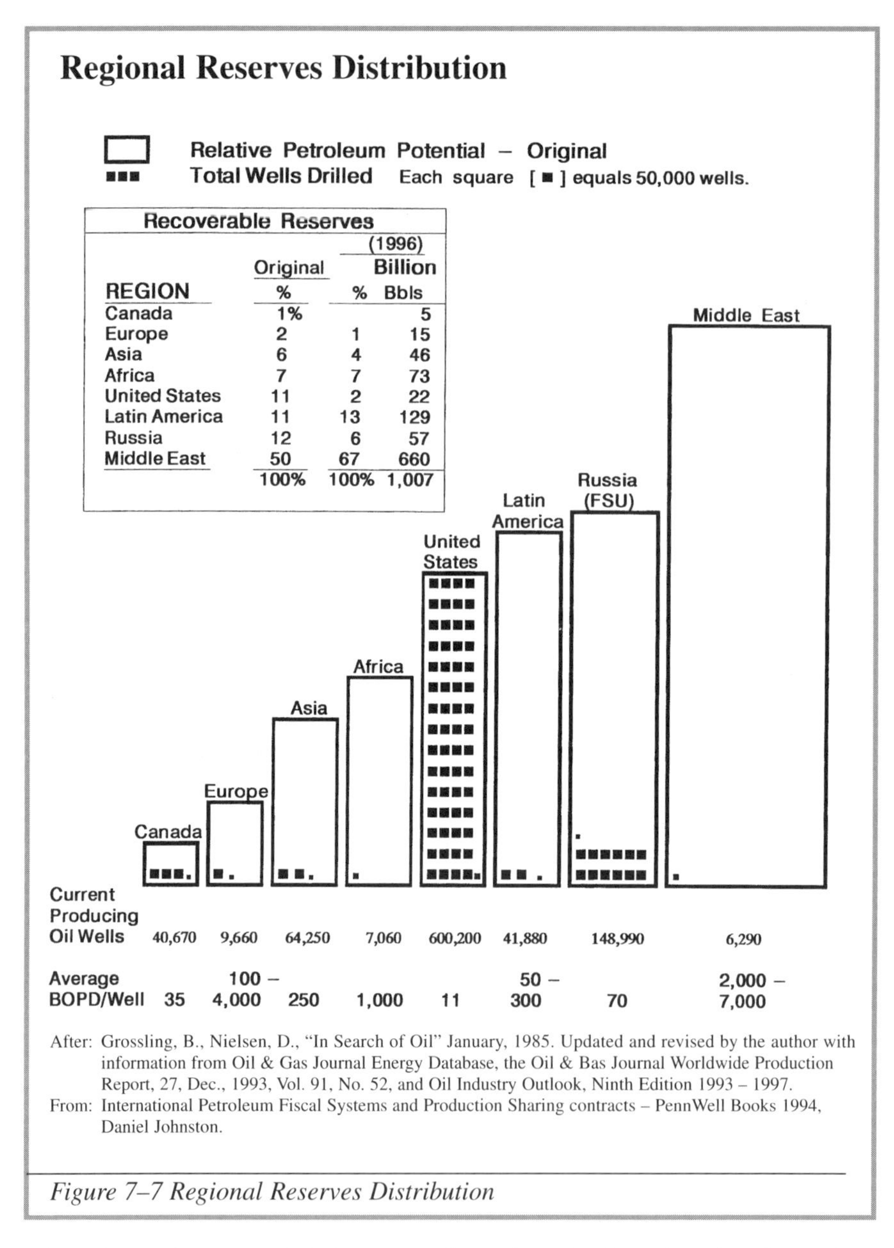

Recoverable Reserves

REGION	Original %	(1996) %	(1996) Billion Bbls
Canada	1%		5
Europe	2	1	15
Asia	6	4	46
Africa	7	7	73
United States	11	2	22
Latin America	11	13	129
Russia	12	6	57
Middle East	50	67	660
	100%	100%	1,007

Figure 7–7 Regional Reserves Distribution

oil per day each) are excluded, the average "lower 48" oil well produces less than 10 barrels of oil per day. And this is an interesting statistic, too.

The U.S. is unique with its stripper well category of oil wells—those that produce an average of less than 10 barrels per day. Nearly 75% of the oil wells in the U.S. are stripper wells. Of these 450,000 odd wells, the average production rate is around two barrels

Area	Wide variation. Average: ±150,000-350,000 acres.
Duration	Exploration: Multiphase, 2–4 years initial + extensions. Production from 20–25 years from startup.
Relinquishment	At end of each main exploration phase, 25% of original area. After end of exploration phase some governments require total relinquishment of all but development area plus "halo."
Exploration Obligations	Wide variations in amount and timing.
Bonuses **Rentals**	Most in highly prospective areas as an equalizer. Usually amount to very little.
Royalty	Most countries have a royalty of some sort. World average is 7%. General range is 0–20%.
Cost Recovery Limit	A phenomenon of PSCs, but about 20% of PSCs do not have a limit. World average is 60–65%. General range is 50–70%.
Depreciation **G&A Expenses**	Not required for cost recovery in 45% of PSCs. World average depreciation is 5-year SLD. Usually a formula based on Capex and Opex (if allowed).
Profit Oil Split	PSC phenomena about 80% are sliding scale. Most are based upon tranches of production.
Taxation	Almost all systems have direct or indirect taxes. Equivalent of corporate income tax. Averages 35%. Others may include withholding taxes at ±15%.
Ringfencing	60% of the countries in the world are ringfenced. 20% have a modified ringfence with some relief. 20% have no ringfence (most likely royalty/tax systems).
Domestic Market Obligation	Fairly rare. Usually a formula that dictates a percentage of entitlement that must be sold to the government at a reduced rate from world price.
Government Participation	About 40% of governments have the option to back-in at discovery/commerciality. Percentage back-in ranges from 10–15% to ±50%. For the countries that have the option, the average is around 30%
Dispute Resolution	Mostly international arbitration.
Other	There is always something else! (Written or unwritten, see Crypto Taxes.)

Table 7–7 Typical Ranges of Key Contract Elements

of oil per day. This is unthinkable to the typical foreign oil man.

Figure 7–7 attempts to capture the difference between the U.S. and the rest of the world—it is an issue of maturity. Our industry has drilled around 3.5 million wells here in the U.S. By comparison in the vast provinces of the former Soviet Union (FSU), less than a million wells have been drilled.

	Typical Development Costs* ($/BBL)
Gulf of Mexico Deepwater	$3.75–4.50
UK North Sea	$4.00+
Southeast Asia	$2.50–3.00

*For economic development

Table 7–8 Typical Development Costs

Crypto Taxes

A tax is a compulsory payment pursuant to the authority of a foreign government. Fines, penalties, interest and customs duties are not taxes. Crypto taxes do not seem to fit the ordinary definitions of taxes, royalties, imposts, duties, excises, and severances, etc. They are almost never captured adequately in the "take" statistics. A few examples include:

- Data purchase costs/fees
- Local office requirement
- Surface rentals
- Training fees and scholarship funds
- Customs duties or customs exemptions that don't hold up
- Cumbersome visa requirements
- Social sphere development costs (written or unwritten)
- Domestic market obligations (DMO)
- Mandatory currency conversions
- Unusually low procurement limitations
- Hiring requirements
- Hostile audits
- Government cost recovery
- Excessive (government owned) pipeline tariffs
- Price cap formulas
- Short loss carry forward periods
- Performance bonds
- Value Added Taxes (VAT) or Goods and Services Taxes (GST)

	1996–1997 Discovery Well Test Rates		Typical Exploration Well Cost	Typical State Take
	Oil (BOPD)	**Gas (MMCFD)**	**($MM)**	**(%)**
UNITED KINGDOM	8,000	22	$8–10	33%
ASIA				
W. Indonesia	560+	23	$3–4	86%
E. Indonesia	560+	23	$3–4	70%
Vietnam	N/A	N/A	$8–12+	57–88%
Australia				
Offshore	7,600 max	10	$3–6+	61%
Pakistan				
Onshore			$2–4+	50–60%
Offshore				38–44%
LATIN AMERICA				
Venezuela	4,000+	10+	$3–16	90+%
Argentina	750	17	$1.5–4	40+%
W. Colombia	700		.3–.5	80+%
E. Colombia	3,500+		20–30	80+%
Peru				60+%
AFRICA				
N. Africa	4,000	30	$2.5–6	Various
W. Africa	6,000		$7–15+	Various
GULF OF MEXICO				
Deep water	800+	15+	$5–13	40%
Shallow water			$3–4	43–48%

Table 7–9 Selected Hot Spots Worldwide

- Reinvestment obligations
- Asset based taxes (Ad Valorem)
- Inefficient allocation mechanisms. Slow indecisive awards
- Unrealistic permitting and impact statements
- Oppressive government controls
- Contract official language other than English

Costs

Just as the geopotential is often an order of magnitude greater in the foreign sector so are the costs. The cost of doing business in the international sector goes beyond the higher cost of drilling even without mobilization and demobilization costs factored in. This is a simple fact of life. There are numerous situations where a simple meeting will require someone with authority to appear in the host country capital. A one-hour meeting overseas can often cost a man-week. And unfortunately, as is often the case when this happens, the meeting may not necessarily resolve any particular important issue.

As a matter of comparison, typical drilling and development costs are summarized in Tables 7–8 and 7–9. Table 7-10 shows well test rates from 186 discoveries worldwide. A comparison of internationally-used terminology for division of profits is shown in Table 7-11.

	1996–1997		
	Lower Quartile	**Average**	**Upper Quartile**
BOPD	840	5,700	10,400
MMCFD	4	23	50

Table 7–10 Well Test Rates from 186 Discoveries Worldwide

The Division of Profits

Government	Company/Contractor
Government equity split Government after-tax equity split (Almost obsolete terminology)	Contractor equity split Contractor after-tax equity split
Government take State take	Contractor take Contractor take
Government marginal take Net government take on the marginal barrel	Contractor take Net contractor take on the marginal barrel
Tax take (excludes govt. participation) Government take (includes govt. participation)	Contractor take Contractor take
Discounted government take*	Discounted contractor take
Government profit share Net cash margin Bottom line financial split-bottom line income split Net net	Contractor Profit Share Net cash margin Bottom line financial split-bottom line income split Net net

*The picture changes dramatically from present value point of view (in favor of government), nevertheless the take statistics undiscounted are very valuable/useful.

Table 7–11 Variations in Terminology

Working Capital Management

8

Short-term management of working capital is an important function. Although working capital management is generic across all organizations and locations, there are unique requirements in the petroleum industry. Significant cash requirements exist for MNOCs and domestic companies to finance crude and product inventories. Volatile crude and product prices demand managing petroleum inventories and maintaining strategic inventories to offset supply shortages created by volatile international political situations. External and uncontrollable forces control crude quotas and prices. High technology, expensive equipment, and dispersion of operations throughout the world require unique strategies to reduce the investment of cash in equipment and parts inventories. These issues require expertise and strategies in all oil and gas companies to manage working capital.

The term *working capital* is familiar to every businessperson, although definitions differ depending on who is asked. The credit analyst in the oil industry may prefer the accounting definition—current assets minus current liabilities. An operating supervisor of a drilling team may define working capital as drill stems and bits needed for the next operation. The entrepreneur may define working capital as cash available to cover disbursement checks. The treasurer may define it as how much short-term borrowing or investing is required to stay afloat. These are not really different definitions; they are different focuses.

Current assets consist of cash and other assets or resources commonly identified as those reasonably expected to be realized in cash, sold, or consumed during the normal operating cycle of the business. The operating cycle is the average time between the acquisition of resources and the final receipt of cash from their sale as the culmination of the entity's revenue-generating activities. If the operating cycle is less than a year, one year is the basis for defining current and noncurrent assets. Current assets include cash and cash equivalents, inventories, receivables, trading securities, certain available-for-sale and held-to-maturity securities, and prepaid expenses.

Current liabilities are obligations whose liquidation is reasonably expected to require the use of existing resources properly classifiable as current assets or the creation of other current liabilities. Current liabilities include payables arising from operations directly related to the operating cycle such as those to obtain materials and supplies, accrued wages, salaries, rentals, royalties, taxes, etc.

Working capital is the sum of the various elements comprising a company's current assets, minus the sum of its various short-term liabilities. But, more importantly, working capital is more than the static snapshot of a portion of the balance sheet. Working capital is a dynamic, ever-changing resource that is the cumulative result of the company's operations together with the amount of invested capital in the firm. Beneath the surface of balance sheet accounts, however, is the hum of operations and the ebb and flow of the business. Examination of those operations illuminates the nature of working capital and its significance.

HOW MUCH WORKING CAPITAL IS REQUIRED?

Measuring working capital without assessing the quality of the assets can be grossly misleading. Consider a company whose receivables represent an average of 55 days' sales outstanding (DSO) with payment terms of "net 30 days." The quality of these receivables should be suspect even though the company appears to have substantial working capital.

Before the 1980s the general feeling regarding receivables from oil and gas companies was always very positive. However, during the 80s some oil companies went bankrupt and up-front guarantees were required from oil companies with a lower credit rating. Also, business with non-oil companies requires the normal scrutiny regarding credit risk.

Likewise, two companies in the same business with equal amounts of inventory may nevertheless present strikingly different scenarios. Green Producing Company's inventory is largely crude to fill refinery orders; Blue Refinery Company's inventories are mostly products awaiting marketing purchasing order; Red Marketing Company's inventories are products awaiting purchasing by the consumer. It could be argued that Green's crude inventory value is superior to Red's product inventory due to the type of consumers. However, price uncertainty is dependent upon daily quoted rates in both areas.

Taking a superficial view of the gross balance sheet numbers is also inadequate. Using the same numbers to calculate historic turnover rates and ratios can be equally misleading. Two companies might have similar inventory turnover rates. However, one company's inventory turnover might include rapid turnover of a small portion of the product line, and slow turnover of the remaining products. This would be less desirable than a company having the same average inventory turnover for over all of its products. Also, the same company may have among its receivables past due accounts that are about to be written off. Day's sales outstanding (DSO) ratios would not disclose this problem. DSO and other ratios are merely averages that often mask underlying problems.

WORKING CAPITAL MANAGEMENT

Working capital management refers to the decisions and strategies concerning amounts and types of investment in current assets and amounts and types of financing from current liabilities. The daily activities of the firm result in a flow of cash through the working capital accounts. Working capital management involves the analysis of risk and reward. A high value for the current ratio (current assets/current liabilities) suggests a strong liquidity position and relatively low risk. However, in general, current assets have low rates of return, while current liabilities have relatively high interest costs. Therefore, there is a high cost to having strong liquidity.

"Source of Capital"

Financing of Working Capital

Many current asset and current liability accounts vary directly with sales activity and these accounts are said to be spontaneous. That is, there are no specific decisions to increase these accounts, but the changes will occur spontaneously with changes in sales. The magnitude of the spontaneous changes will have a significant impact on the amount of external financing necessary for working capital. Increases in noncash current assets must be offset by either reducing cash or increasing liabilities or debt.

Decreases in current liabilities must be offset by decreases in an asset account or increases in other liability accounts. A decrease in accounts payable requires either a decrease in cash or an increase in debt to pay off the accounts. A larger volume of sales activity would typically require a higher volume of materials purchased. If these materials are purchased on credit, the accounts payable account will increase.

The total of the above net increases in current asset accounts (excluding net cash) represents the total amount of financing necessary for the incremental working capital, assuming there are no net changes in cash flows from changes in long-term assets or liabilities. If the level of current assets (excluding net cash) decreases, this amount represents funds that have been freed up or no longer need to be financed. Lower inventory or lower accounts receivable under rising sales results in increased liquidity for the firm. Decreases in current liabilities represent a reduction in funding available to support the current assets. The firm would then have to utilize external sources such as long-term debt or other capital to support the working capital investment.

Short-Term Credit

Since short-term credit offers the advantage of greater flexibility, and sometimes lower cost, most firms use at least some current debt in spite of the fact that short-term debt increases the firm's risk. The four major types of short-term credit are (1) accruals, (2) accounts payable or trade credit, (3) bank loans, and (4) commercial paper.

The traditional view holds that permanent assets should be financed with long-term capital, while temporary assets should be financed with short-term credit. Using short-term credit to finance long-term assets involves the risk that the firm will be unable to pay off debts as they come due.

There are advantages and disadvantages to the use of short-term credit. One advantage is flexibility; short-term debt may be repaid if the firm's financing requirements decline. Another advantage is cost; short-term interest rates are normally lower than long-term rates because ordinarily less risk is involved. Therefore, financing with short-term credit usually results in lower interest rates. A disadvantage is that short-term debt is generally more risky than long-term debt for two reasons: (1) short-term interest rates fluctuate widely while long-term rates tend to be more stable and predictable, and (2) short-term debt comes due every few months. If a firm does not have the cash to repay debt when it comes due, and if it cannot refinance the loan, it may be forced into bankruptcy.

Working Capital Cash Flow Cycle

The flow of cash through the working capital accounts can be illustrated with the cash flow cycle. It shows how the strategies concerning inventory, receivables and payables all interact to affect the flow of cash through the firm (Fig. 8–1).

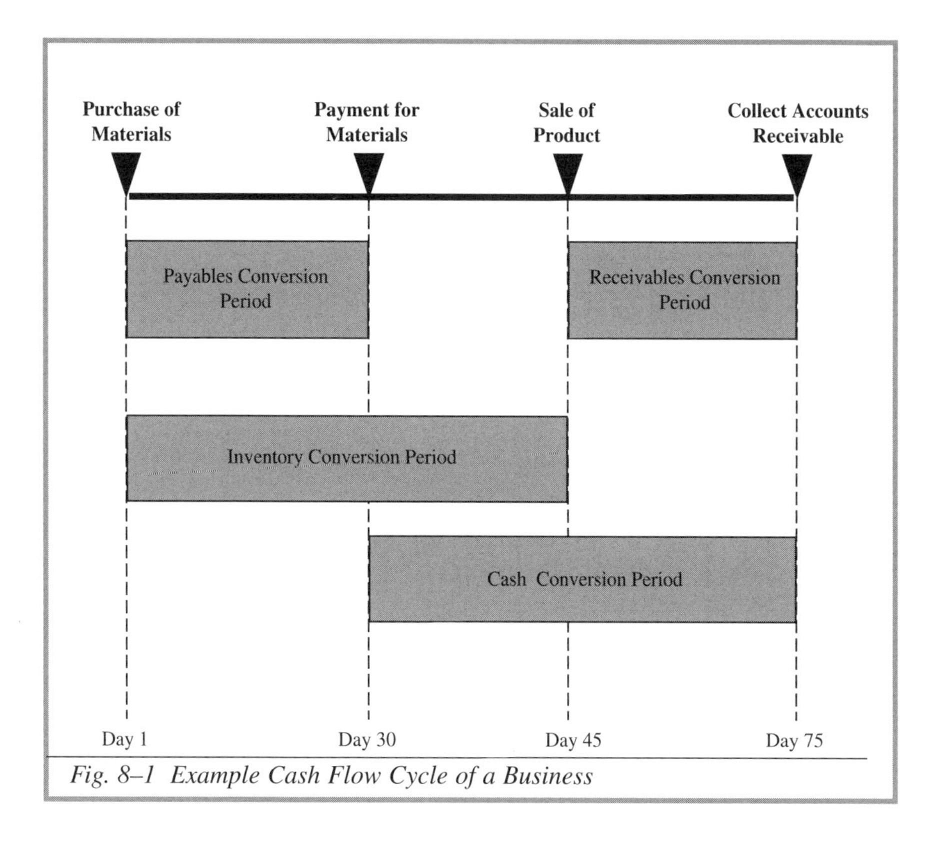

Fig. 8–1 Example Cash Flow Cycle of a Business

PAYABLES CONVERSION PERIOD is the average number of days between the purchase of materials and supplies and payment for them. The ability to purchase materials on deferred payment terms allows the firm to delay the outflow of cash. Any strategy that affects payables, such as stretching payables or taking cash discounts, will affect the payables conversion period. The stretching of payables enables a firm either to use cash to reduce debt or to invest cash on a short-term basis to increase interest income.

Inventory Conversion Period is the average number of days, from the acquisition of crude until the sale of the petroleum products produced from crude, that usually generates accounts receivable. Inventory management techniques such as just in time (JIT) can help to reduce significantly the absolute levels of materials in inventory.

Receivables Conversion Period is the average number of days required to convert the accounts receivable into cash. Any change in accounts receivable policies, such as credit standards, credit terms or collection efforts, will impact the receivables conversion period. The receivables conversion period is also referred to as *day's sales outstanding* (DSO) or *average collection period.*

Cash Conversion Cycle is the average number of days between the actual cash outflow for the acquisition of materials and supplies and the actual cash inflow from the sale of the products

Given the following statements (in millions of dollars), determine the firm's cash conversion cycle.

Cash	$ 100	Accounts Payable	$ 150
Accounts Receivable	150	Notes Payable	100
Inventory	250	Long-term Debt	350
Fixed Assets	500	Owner's Equity	400
Total Assets	$1,000	Total Liabilities and Equity	$1,000

Assume the firm has annual sales of $1,750 and cost of goods sold (COGS) equal to $1,095. Also assume that beginning inventory was $225. (All figures are in millions of dollars)

Purchases	=	Ending Inventory + COGS – Beginning Inventory
	=	$250 + $1,095 – $225 = $1,120
Inventory Conversion Period	=	(Ending Inventory/COGS) x 365
	=	($250/$1,095) x 365 = 83.3 days
Receivables Conversion Period	=	(Accounts Receivable/Sales) x 365
	=	($150/$1,750) x 365 = 31.3 days
Payables Conversion Period	=	(Accounts Payable/Purchases) x 365
	=	$150/$1,120) x 365 = 48.9 days
Cash Conversion Cycle	=	Inventory Conversion Period
	+	Receivables Conversion Period
	–	Payables Conversion Period
	=	83.3 days + 31.3 days – 48.9 days
	=	65.7 days

Figure 8–2 Cash Conversion Cycle

or services. The cash conversion cycle can be expressed by the formula and calculations in Figure 8–2. In the calculation of the cash conversion cycle, each of the periods is based on annualized information. When used for a shorter period (such as a quarter) the COGS, sales, and purchases figures must be adjusted to reflect data only for the quarter. Similarly, quarter-end rather than year-end information must be used for inventory, accounts receivable, and accounts payable. Finally, rather than multiplying by 365, the number of days in the quarter should be used.

Using the information from Figure 8–2 example, the firm must finance approximately 66 days of sales. Assuming a cost of capital of 12%, the cost of the conversion cycle is:

Cost of the
Conversion Cycle = 66 days x Average Daily Sales x (Cost of Capital of 12%)
= 66 x (1,750,000,000/365) x .12
= $37,792,603

Figure 8–3 Cost of Cash Conversion Cycle

The cash conversion cycle tells the firm how long, on average, it must finance the cash outflow before it receives cash inflow. This cycle can be considered in a present value framework. The treasurer should examine the timing of cash inflows and outflows as they relate to the conversion of raw materials into the receipt of available funds from sales. The cash conversion cycle can be shortened by reducing the inventory conversion period, by reducing the receivables conversion period or by lengthening the payables deferral period. The firm may be able to improve profits by reducing financing costs if it can shorten the cash conversion cycle.

Using the number of days calculated in Figure 8–2, the firm must finance approximately 66 days of sales at a cost of capital of 12%. The cost of cash conversion as calculated in Figure 8–3 is $37.8 million. This represents the cost of financing the net investment in inventory and accounts receivable less accounts payable during the year.

Financing Current Assets

The firm's need for current assets will typically fluctuate with variations in its sales. Any firm will have a sales cycle with major sales fluctuating and falling below some minimum level during the cycle. There is usually a need for a base level of current assets at all times during the sales cycle. This minimum required current assets investment is generally referred to as the firm's permanent current assets.

A firm may also decide to try to smooth out its use of short-term debt over time, borrowing on a regular basis whether or not it needs funds at a particular time. This approach may yield more stable borrowing rates, especially if more favorable rates from the bank can be negotiated.

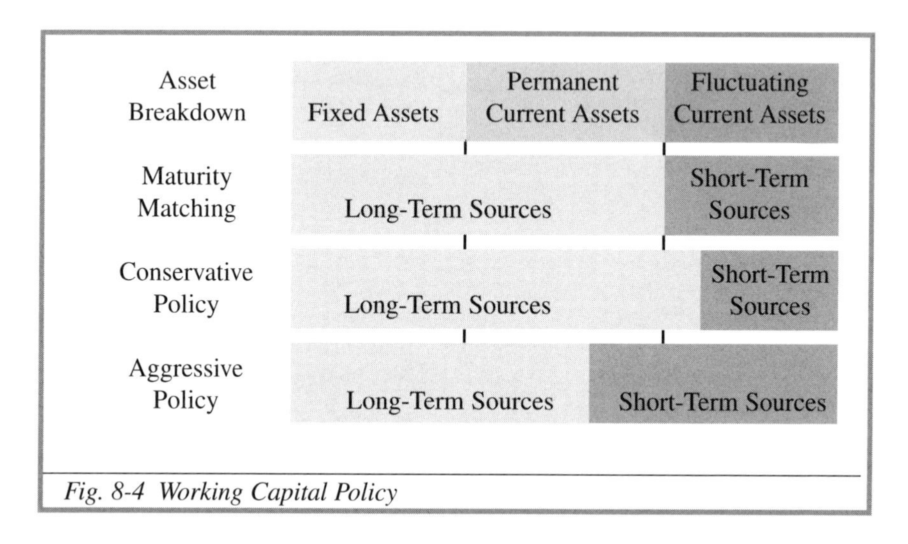

Fig. 8-4 Working Capital Policy

Maturity Matching. The firm's total assets are defined as its fixed assets plus permanent and temporary current assets. Under maturity matching, the total permanent assets are financed with long-term financing. Short-term financing is used to finance the temporary current assets. The need for short-term financing follows exactly the pattern of fluctuation of the temporary current assets. This is generally considered a traditional or moderate policy.

Aggressive Strategy. Under the aggressive approach, the firm finances all of its fixed assets, but it finances only a portion of its permanent current assets with long-term financing. Short-term financing is used for the remainder of the permanent current assets and for all of the temporary current assets. In comparison with other policies, this policy uses a larger amount of ongoing short-term financing. Since short-term financing often costs less than long-term financing, this strategy is usually less costly. However, greater use of short-term financing will result in a lower current ratio and greater risk. This risk arises from both interest rate risk (changing rates) and the risk that the loan may not be renewed.

Conservative Strategy. Under the conservative approach, long-term financing is used for the fixed assets, the permanent current assets and some portion of the temporary current assets. Generally, the treasurer will try to finance the average level of temporary current assets with long-term sources. Short-term financing is used for the remainder of the temporary current assets. The relatively smaller use of short-term financing will result in a larger current ratio, but may also result in lower profits for the firm. The risk/return trade-off must be analyzed in both current asset investment and current asset financing decisions. The use of derivative financing products may allow the treasurer to shift or hedge some of the risk from an aggressive investment policy.

CASH MANAGEMENT

The term *cash management* can be broadly defined to mean optimization of cash flows and investment of excess cash. Firms hold cash for four primary reasons:

1. *Transactions*. Cash balances are necessary to conduct business. Payments must be made in cash and receipts are deposited in the cash account. Cash balances associated with routine payments and collections are known as transactions balances.
2. *Precaution*. Cash inflows and outflows are somewhat unpredictable, with the degree of predictability varying among firms and industries. Therefore, just as firms hold safety stocks of inventories, they also need to hold some cash in reserve for random, unforeseen fluctuations in inflows and outflows. These amounts are defined as precautionary balances.
3. *Speculation*. Some cash balances may be held to enable the firm to take advantage of any bargain purchases that might arise; these are defined as speculative balances. However, as with precautionary balances, firms today are more likely to rely on reserve borrowing power and on marketable securities portfolios than on actual cash holdings for speculative purposes.
4. *Compensation to banks for providing loans and services*. Banks provide services to firms by clearing checks, operating lockbox plans, supply credit information, and the like. These services cost the bank money, so the bank must be compensated for rendering them. Compensating balances therefore provide banks funds for lending purposes. Firms generally push for centralized banking policies to minimize the amount of compensating balances outstanding and for better control over deposited amounts.

A *lockbox plan* is used to reduce mail and clearing delays. Suppose a firm located in a central location makes sales to customers all across the country. It can arrange to have its customers send payments to post office boxes (lockboxes) in their own local areas. A local bank will pick up the checks, have them cleared in the local areas, and then transfer the funds to the company headquarter's bank. In this way collection time can be reduced by one to five days. Examples of freeing funds in the amount of $5 million or more by this method are not uncommon.

Cash in the cash register is a nonearning asset. Excessive cash balances lower asset turnover, which reduces the rate of return on net worth and the value of the stock. Large companies manage cash daily to avoid idle cash. Petty cash funds have been eliminated and credit cards have replaced employee advances for travel and entertainment expenses.

Speeding collections and slowing payments permit a firm to operate with smaller cash balances. Companies frequently use a lockbox plan to reduce the time required to process checks and thus to receive payments. Electronic funds transfers also speed collections and make funds available more rapidly. Slowing payments improves efficiency by keeping cash on hand for longer periods of time. *Float* is the difference between the balance shown in a firm's checkbook and the balance on the bank's books. Positive float is an indication that the firm is more efficient in making collections than are the recipients of its checks.

Establishing the Cash Balance

The most important tool in cash management is the cash budget or forecast showing the firm's projected cash inflows and outflows over a specified period of time. Cash forecasts can be constructed on a monthly, weekly, or even daily basis. The procedures for establishing a cash forecast can be simple or complex, but the objective is to provide cash requirements and balances. The basic purpose behind the preparation of the cash forecast is to plan so that the business will have the necessary cash—whether from the short-term or long-term viewpoint. Further, when excess cash is to be available, forecasting provides a means of anticipating an opportunity for effective utilization. The cash forecast can be used to:

- Point out peaks or seasonal fluctuations in business activity that necessitate larger investments in inventories and receivables.
- Indicate the time and extent of funds needed to meet maturing obligations, tax payments, and dividend or interest payments.
- Assist in planning for growth, including the required funds for plant expansion and working capital.
- Indicate—well in advance of needs—the extent and duration of funds required from outside sources, thus permit the securing of more advantageous loans.

- Assist in securing credit from banks and improve the general credit position of the business.
- Determine the extent and probable duration of funds available for investment.
- Plan the reduction of bonded indebtedness or other loans.
- Coordinate the financial needs of the subsidiaries and divisions of the company.
- Permit the company to take advantage of cash discounts and forward purchasing, thereby increasing its earnings.

Sources of Cash Receipts

The sources of cash receipts for the typical firm are well-known: collections on account, cash sales, royalties, rent, dividends, sale of capital items, sale of investments, and new financing. Usually, the most important recurring sources are collections on account and cash sales. Experience and knowledge of trends will indicate the probable share of total cash sales. From the sales forecast, then, the total cash sales value can be determined. In a somewhat similar fashion, information based on experience can be gleaned from the records to provide an estimate of collection of credit sales (Fig. 8–5).

These experience factors must be modified, not only by trends developed over a period of time, but also by the estimate of general business conditions as reflected in collections, as well as contemplated changes in terms of sale or other credit policies. Refinements in the approach can be made if experience varies widely between geographical territories, types of customers, or channels of distribution.

	JAN.	FEB.	MAR.
Sales (net of cash discounts)	$ 5,000	$ 5,000	$10,000
Collections:			
During month of sale (20%)	1,000	1,000	2,000
During first month after sale (70%)		3,500	3,500
During second month after sale (10%)			500
Total Collections	$ 1,000	$4,500	$6,000
Purchases (70% of next mos. sales)	$ 3,500	$7,000	$10,500
Payments (one-month lag)		$3,500	$ 7,000

Fig. 8–5 Calculation of Cash Collections and Disbursements

Estimating Cash Disbursements

The operating budgets generally provide a basis for estimating cash disbursements for the labor budget, material budget, operating expense budget, and capital budgets. Again, the cash disbursement forecasts will need to be adjusted over time, rising and falling with seasonal patterns and with longer term changes in the scale of the firm's operations (Fig. 8–6).

	MARCH
Collections	$6,000
Payments	
Purchases	$ 7,000
Wages and Salaries	750
Rent	250
Other Expenses	100
Total Payments	$8,100
Net Cash Gain (loss) during month	
Collections—Payments	($2,100)
Cash at start of month if no borrowing is done	3,000
Cumulative Cash = (cash at start + gains and/or – losses)	$ 900
Deduct Target Level of Cash	2,500
Total Cash Borrowing to Maintain $2,500 Target Cash Balance	$ 1,600

Fig. 8–6 Calculation of Cash Forecast

Simple Forecasting Formula

Although the forecast of cash requirements can be made by using the worksheet method, it is often easier to use a simple forecasting formula. In addition, the formula can be used to clearly show the relationship between sales growth and financing requirements.

The equation must be used with caution if excess capacity exists in any of the asset accounts. The equation can also be used to calculate the maximum growth rate that can be financed without external funds. This is done by setting the additional funds requirement equal to zero and solving for the growth rate in sales. If management cannot or does not wish to use external financing, it may have to limit potential growth and turn away sales beyond some specified amount. The Simple Forecasting Formula is demonstrated in Figure 8–7.

Simple Cash Forecasting Formula

AFN = (A/S x S2) - (L/S x S2) - [M x Sl (1 – D)]

where

AFN	=	Additional funds needed
A/S	=	Assets that increase spontaneously with sales as a percentage of sales. Or required dollar increase in assets per $1 increase in sales: 75%
S0	=	Last year's sales: $400,000
S1	=	Total expected sales for the year in question: $600,000
S2	=	Change in sales (S1– S0) $600,000 – $400,000 = $200,000
L/S	=	Liabilities that increase spontaneously with sales as a percentage of sales, or spontaneously generated financing per $1 increase in sales: 22.5%
M	=	Profit margin or rate of profits after taxes per $1 of sales:10%
D	=	Percentage of earnings paid out in dividends: 60%

Substituting

AFN = [.75 ($600,000 – $400,000)] – [.225 ($200,000)] – [.1($600,000 x (1 – .60)]

AFN = $150,000 – $45,000 – $24,000

AFN = $81,000 additional funding required

Fig. 8–7 Formula for Forecasting Cash Requirements

The resulting $81,000 additional funding requirement at the expected level of sales and new assets will indicate that corporate policies need to be adjusted, i.e., arrange for new borrowing or decrease sales volume which will automatically reduce the need for new capital projects.

International Cash Management

From an international perspective, cash management in oil and gas companies is very complex because of different laws among countries that pertain to cross-border cash transfers. In addition, exchange rate fluctuations can affect the value of cross-border cash transfers. On the other hand, the oil and gas company's cash management is simplified because crude value is denominated in petrodollars. However, local purchases are denominated in local currency.

The international cash management role may be separated into two functions: (1) optimizing cash flow movements, and (2) optimizing use of available funds. To optimize cash flow movements, the multi-national oil corporation (MNOC) may consider using preauthorized checks or placement of lockboxes to accelerate inflows. In addition, it may use a netting scheme to minimize float and currency conversion costs.

Netting can be implemented with the joint effort of subsidiaries or by the centralized cash management group. This netting scheme technique optimizes cash flow by reducing the administrative and transaction costs that result from currency conversion. Consider an MNOC with two subsidiaries located in different countries. Any time one subsidiary purchases goods from a second subsidiary, it may need a foreign currency to make payment. The second subsidiary may do the same when purchasing goods from the first subsidiary. Both subsidiaries could avoid or at least reduce the transaction costs of currency conversion if they netted out the payments—that is, if they accounted for all of their transactions over a given period to determine one net payment.

To optimize use of available funds, the MNOC may develop a centralized cash management strategy whereby excess funds at the individual subsidiaries are pooled. This may allow for higher yields due to larger deposits. A centralized approach can also accommodate short-term financing needs of some subsidiaries with excess funds at other subsidiaries.

Also, the MNOC may be willing to set up a foreign currency deposit. This may be a worthwhile strategy if the expected effective yield on such a strategy is sufficiently above the return on a domestic deposit to outweigh the exchange rate risk. To forecast the effective yield, a forecast of the exchange rate reflecting the deposit's currency is necessary. For this purpose, the MNOC can develop either a point forecast or a probability distribution. Once the exchange rate forecast is used to determine the expected effective yield on the foreign deposit, the MNOC can compare its risk and potential reward characteristics with a domestic deposit. Even if several MNOCs are faced with the same decision and have a similar assessment, their actual choice might vary since their degree of risk aversion might vary.

As a final alternative, the MNOC should consider investing in a portfolio of foreign currency deposits expected to generate high effective yields. This can often lead to a higher probability that the overall portfolio yield would be more rewarding than a domestic investment, especially if the movements of these currencies are not highly correlated.

ACCOUNTS RECEIVABLES

The typical firm has more than 20% of its total assets invested in accounts receivable, and effective management of these assets is important to the profitability and risk level of the firm.

The total amount of accounts receivable outstanding is determined by (1) the volume of credit sales, and (2) the average length of time between sales and collections.

Sales and collections are influenced by a firm's credit policy which includes four variables:

1. The *credit period* is the length of time for which credit is granted. Increasing the credit period often stimulates sales, but there is a cost involved in carrying the increased receivables. The credit period varies for the type of customer. In the upstream sector of the oil industry, the credit term for crude is generally 30-60 days depending upon distance between source, type of customer, and type of contract. In the downstream sector of the oil industry, sale of LPG to farmers may be established at "after harvest," and sale of asphalt to highway contractors will depend upon the payment policy of the government agency.
2. *Credit standards* relate to decisions about who will be granted credit. If the firm extended credit sales to only the strongest of customers, it might never have bad debt losses. On the other hand, it would probably be losing sales.
3. *Collection policy* refers to the procedures the firm follows to collect past-due accounts. The collection process can be expensive in terms of both direct costs and lost goodwill, but at least some firmness is needed to prevent an undue lengthening of the collection period and to minimize outright losses.
4. *Cash discounts* are the last variable in the credit policy decision. Cash discounts attract customers and encourage early payment but reduce the dollar amount received on each discount sale.

Firms use *incremental analysis* to analyze increases or decreases in sales and costs associated with changes in one or more of the credit policy variables (Fig. 8–8). The difference between incremental sales and incremental costs is defined as *incremental profit*. Changes in credit policy will affect one or more of the following variables: sales, production costs, bad debt losses, discount expenses, level of accounts receivable, costs of administration, and collection expense.

The effectiveness of a firm's credit policy can be measured by examining its accounts receivable turnover ratio and its average collection period. Trends may be evaluated or results compared with industry averages. Procedures in management of accounts receivable are:

- An aging schedule of accounts receivable breaks down accounts according to how long they have been outstanding.
- Sometimes a carrying charge is assessed on accounts receivable. This is an offset to the cost of carrying the receivables.
- Some companies finance their receivables by selling their accounts at discount rates, generally without recourse, to a financial institution.

Incremental Investment in Receivables

CI = [(ACPI – ACPO) x (S0/365)] + [V(ACPI) (S1/365)]

Incremental Change In Profits

CP = [S1 (I – V)] – [k(CI)]

where

CI	=	Incremental change in the level of the firm's Investment in Accounts Receivable
CP	=	Incremental change in profit
S0	=	Current Sales = $100,000
S1	=	Incremental Sales = ($150,000 – $100,000) = $50,000
S2	=	New sales total = $150,000
V	=	Variable costs as a percent of sales = 60%
CM	=	Contribution Margin = 100% – 60% = 40%
k	=	Cost of Capital = 10%
ACPO	=	Average collection period prior to a change in credit policy = 0
ACPI	=	New average collection period after the credit policy change = 30

Substituting in equation

Incremental Investment in Receivables

$$CI = \left[(30 - 0) \times \frac{\$100{,}000}{365}\right] + \left[(60\% \times 30) \times \frac{\$50{,}000}{365}\right]$$

CI = $8,219 + $2,466 = **$10,685 Additional Investment in AR**

Incremental Change in Profits

CP = [50,000 x (1 – 60)] – (.10 x $10,685)

CP = $20,000 – $1,069 = **$18,431 Additional Profit from Increased Sales**

Fig. 8–8 Incremental Analysis of Change in Credit Policy upon Investment in AR and Increased Profit from Increased Sales

Both consumer and commercial accounts exist in the oil and gas industry. The *consumer accounts* represent the end users at the gasoline station, product terminals, or the burner tips. Marketing segments in the oil and gas industry have developed various techniques for managing consumer credit. Most companies use some form of credit cards whether issued by the company or banks. In recent years, oil companies have outsourced their consumer credit to banks through use of Visa or Master Cards. When the companies issue credit cards, a credit card division is utilized to handle billing, credit, and collection. In addition, the company must finance the outstanding receivable.

The *commercial accounts* represent utility plants using gas, airlines using jet fuel, railroads using diesel fuel, road contractors using asphalt, etc. In addition, there are other types of inter-industry accounts such as refineries buying crude, marketing companies buying products, etc. Each classification represents a different type of customer with unique credit requirements.

Special credit policies are designed to meet the needs of the customers and provide a cash stream to the seller. Letters of credit issued by a bank for a customer or prepayments are sometimes used for high-risk accounts.

Accounts Receivable Financing

Some of the more popular methods of financing both international and domestic trade receivables include the following seven methods:

1. In *accounts receivable financing* the seller or exporter may require financing from a bank in which the bank will provide a loan to the seller secured by an assignment of the accounts receivable. Most banks establish advance ratios ranging from 60-80 % of the invoice amount. It is important to note that the bank's loan is made to the exporter based on its creditworthiness. In the event the buyer fails to pay the exporter for whatever reason, the exporter is still responsible to repay the bank.
2. *Factoring* is a type of financing in which the seller or exporter sells the accounts receivable without recourse. The factor then assumes all administrative responsibilities involved in collecting from the buyer and the associated credit exposure. Because the importer must be creditworthy from a factor's point of view, crossborder factoring is often used. This involves a network of factors in various countries that assess credit risk. The exporter's factor contacts a correspondent factor in the buyer's country to assess the importer's creditworthiness and handle the collections of the receivable.
3. *Letter of credit* (L/C) is a critical component of many international crude oil transactions. The L/C is an undertaking by a bank to make payments on behalf of a specified party to a beneficiary under specified conditions. The beneficiary (exporter) is paid upon presentation of the required documents in compliance with the terms of the L/C. Trade-related letters of credit are known as commercial L/Cs or import/export letters of credit. There are basically two types: A revocable L/C can be cancelled at any time without prior notification to the beneficiary, and it is seldom used. An irrevocable L/C cannot be cancelled or amended without beneficiary approval.
4. *Draft,* also known as *bill of exchange*, is an unconditional promise drawn by one party, usually the exporter, on the importer or importer's bank (drawee), requesting the drawee to pay the face amount of the draft at sight or at a specified future date. If it is payable at a specified

future date, and the importer accepts it, it is known as a *trade acceptance*. A *banker's acceptance* is a time draft drawn on and accepted by a bank.

5. *Short-term bank loans*, as previously explained, can allow an exporter to receive funds immediately, yet delay payment until a future date. The bank may even provide short-term loans beyond the banker's acceptance period. In the case of an importer, the purchase from overseas usually represents the acquisition of inventory.
6. *Countertrade* denotes all types of foreign trade transactions in which the sale of goods to one country is linked to the purchase or exchange of goods from that same country.
7. *Barter* represents the exchange of goods between two parties without the use of any currency as a medium of exchange. Many multinationals have been confronted with countertrade opportunities, particularly in Asia, Latin American and Eastern Europe. In a *compensation arrangement*, the delivery of goods to one party is compensated for by buying back a certain amount of product from that same party. The term *counter purchase* denotes the exchange of goods between two parties under two distinct contracts expressed in monetary terms. Delivery and payment of both goods are technically separate transactions.

MANAGEMENT OF ACCOUNTS PAYABLE AND ACCRUALS

Accounts payable and accrued expenses arise from delayed payments. Accounts payable come from the use of other production inputs, such as labor, materials and administrative costs for which the payment is delayed. The total amount of financing provided by these sources is a function of the level of activity, over which the financial manager may have little influence, and the length of the time deferral on the payment.

Payables and accruals, generated through the normal operating activities of a firm rise and fall spontaneously with the level of sales. As the amount of materials and inventory used in production increases or inventories increase, the financing provided by payables and accruals increases in the short term. This reduces the need for external financing as long as the inventory-holding period does not substantially exceed payment terms.

Accounts Payable Decisions

These decisions include specifying the credit terms (where negotiation with the seller or vendor is possible) and determining when to pay. If the buyer is an important customer, the seller may be willing to negotiate payment terms. The terms may be affected by the quantity or frequency of purchase and the form of payment (e.g., electronic versus paper check). Some payment terms contain options for different payment (discounts) conditional upon when the payment is made. Typical discount terms might be: 2/10, Net 30 (a 2% discount may be taken if the payment is made in 10 days, otherwise the full amount is due in 30 days). The payor should take the payment option that minimizes the net present value cost of the payment. Any direct and indirect penalty costs that may result from delaying the payables beyond the due date should be considered. Many small companies may have no choice other than taking the credit terms offered, as they have no access to other sources of capital.

If a firm forgoes a cash discount but pays the bill on the final date of the net the firm has incurred an *opportunity cost*. For example, if the terms of sale are 2/10, net 30 the firm has the use of the funds for an additional 20 days if it does not take the cash discount but pays on the final day of the net period.

Cost of Not Taking Cash Discount

Assuming credit terms of 2/10, Net 90, and your cost for short-term funds is 8%, should you take cash discount?

$$\text{Effective Cost of Discount (i)} = d/(1 - d) \times 365/(n - t)$$

where

d	=	cash discount (2%)	=	.02
n	=	number of days for net	=	90
t	=	number of days for discount	=	10

$$\text{Effective Cost of Discount} = .02/(1 - .02) \times 365/(90 - 10) = 9.31\%$$

Fig. 8–9 Opportunity Cost of Not taking Discounts on Purchases

Compare the cost of not taking the discount to the short-term cost of capital or the opportunity cost for investing short-term funds. In Figure 8–9, the company should take the discount, as the cost of not taking the discount is 9.31% versus the firm's short-term borrowing costs of 8%. The firm would, therefore, borrow funds at 8% and take the discount. If, however, the firm's short-term cost of funds was 10%, it should not take the discount but pay the net amount on Day 90.

From the seller's point of view, the benefit of offering a discount is receiving the funds earlier, while the cost is the discount given to the buyer in order to encourage early payment.

As indicated in Figure 8–10, the seller is better off by $170 per payment if the buyer takes the discount and pays on Day 10, rather than paying the full amount on Day 90.

Assume that the seller has a cost of capital of 15%, the average credit sale is $100,000, and the offered terms are 3/10, Net 90.

The benefit of receiving the discounted payment on Day 10 is:

Cost (10) = [$100,000 x (1 – .03)] / [1 + (10/365) x .15] = $96,603

The value of receiving the full payment (PV) on Day 90 is:

Cost (90) = [$100,000] / [1 + (90 / 365) x .15 = $96,433

Net Benefit = PV10 – PV90 = $96,603 – $96,433 = $170

Fig. 8–10 Benefits of Cash Discount to Seller

Borrowing on an unsecured basis is generally cheaper and simpler than on a secured loan basis because of the administrative costs associated with the use of security. However, lenders will demand some form of collateral if a borrower's credit standing is questionable. Most secured short-term business loans involve the pledge of short-term assets such as accounts receivable or inventories. The legal procedures for establishing loan security have been standardized and simplified in the Uniform Commercial Code.

Terms of Payment for International Trade

In any international trade transaction, credit is provided by either the supplier (exporter), buyer (importer), one or more financial institutions, or any combination of these. The supplier may have sufficient cash flow to fund the entire trade cycle, beginning with the production of the product until payment is eventually made by the buyer. This is known as supplier credit. In some cases, the exporter may require bank financing to augment its cash flow. On the other hand, the supplier may not desire to provide financing, in which case the buyer will have to finance the transaction itself, either internally or externally with its bank. Banks on both sides of the transaction can thus play an integral part in trade financing.

In general, five basic methods of payment are used to settle international transactions, each varying in the degree of risk to the exporter and importer:

1. Under the *prepayment* method, the exporter will not ship the goods until the buyer has remitted payment to the exporter. Payment is usually made in the form of an international wire transfer to the exporter's bank account or bank draft. This method affords the supplier the greatest degree of protection, and it is normally requested of first-time buyers whose credit worthiness is unknown or countries in financial difficulty. Most buyers, however, are not willing to bear all the risk by prepaying an order.

2. A *letter of credit* (L/C) is an instrument issued by a bank on behalf of the importer (buyer) promising to pay the exporter (beneficiary) upon presentation of shipping documents in compliance with the terms stipulated therein. In effect, the bank is substituting its credit for that of the buyer. This method is a compromise between seller and buyer because it affords certain advantages to both parties. The exporter is assured of receiving payments from the issuing bank as long as it presents documents in accordance with the L/C. On the other hand, the importer does not have to pay for the goods until shipment has been made and documents are presented in good order.
3. A *draft* (or bill of exchange) is an unconditional promise drawn by one party, usually the exporter, on the buyer, instructing the buyer to pay the face amount of the draft upon presentation. The draft represents the seller's formal demand for payment from the buyer. A draft affords the supplier less protection than a L/C since the banks are not obligated to honor payments on the buyer's behalf.
4. Under a *consignment arrangement*, the exporter ships the goods to the importer while still retaining actual title to the merchandise. The importer has access to the inventory but does not have to pay for the goods until they have been sold to a third party.
5. The opposite of prepayment is the *open-account transaction* in which the exporter ships the merchandise and expects the buyer to remit payment according to the agreed-upon terms. The exporter is relying fully upon the financial credit-worthiness, integrity, and reputation of the buycr. As might be expected, this method is used when seller and buyer have a great deal of experience with each other and mutual trust.

International Terms of Payment and Currency Rate

Before the MNOC's parent or subsidiary needing funds searches for outside funding, it should determine whether there are internal funds available. That is, it should check other subsidiaries' cash flows. If, for example, earnings have been high at a particular subsidiary and a portion of funds generated is simply invested locally in money market securities, the parent may request these funds from the subsidiary. This is especially feasible during periods when the cost of obtaining funds in the parent's home country is relatively high.

Regardless of whether an MNOC parent or subsidiary decides to obtain financing from subsidiaries or from some other source, it also must decide which currency to borrow. Even if it needs its home currency, it may prefer to borrow a foreign currency to offset a net receivables position in that foreign currency or to reduce borrowing costs if the interest rates on such currencies are attractive.

For example, consider an U.S. firm that has net receivables denominated in German marks. If it needs short-term funds, it could borrow marks and convert them to U.S. dollars. Then the net receivables in marks will be used to pay off the loan. In this example, financing in a foreign currency reduces the firm's exposure to fluctuating exchange rates. This strategy is especially appealing if the interest rate of the foreign currency is low. Even when an MNOC parent or subsidiary is not attempting to cover foreign receivables, it may still consider borrowing foreign currencies if the interest rates on such currencies are attractive.

Financing in foreign currencies is common as a result of the development of the Eurocurrency market. In reality, the value of the currency borrowed will most likely change with respect to the borrower's local currency over time. The actual cost of financing by the debtor firm will depend on (1) the interest rate charged by the bank that provided the loan, and (2) the movement in the borrowed currency's value over the life of the loan. Thus, the actual or "effective financing rate" may differ from the quoted interest rate. This point is illustrated in the following example (Fig. 8–11).

A U.S. firm is given a one-year loan of 1,000,000 Swiss francs at the quoted interest rate of 8%. When the U.S. firm receives the loan, it converts the Swiss francs to U.S. dollars to pay a supplier for materials. The exchange rate at the time is $.50 per Swiss franc, so the 1,000,000 Swiss francs are converted to $500,000 (computed as 1,000,000 francs x $.50 = $500,000). One year later, the U.S. firm pays back the loan of 1,000,000 Swiss francs plus interest of 80,000 Swiss francs (interest computed as 8% x 1,000,000 Swiss francs.) Thus the total amount of Swiss francs (SF) needed by the U.S. firm is SF 1,000,000 + SF 80,000 = SF 1,080,000 francs. Assume the Swiss franc appreciates from $.50 to $.60 by the time the loan is to be repaid. The firm will need to convert $648,000 (computed as 1,080,000 francs x $.60 per franc) to secure the necessary number of francs for loan repayment.

To compute the effective financing rate, first determine the number of dollars beyond the amount borrowed that were paid back. Then decide by the number of dollars borrowed (after converting the francs to dollars). Given that the firm borrowed the equivalent of $500,000 and paid back $648,000 for the loan, the effective financing rate in this case is $148,000/$500,000 = 29.6%. If the exchange rate remained constant throughout the life of the loan, the total loan repayment would have been $540,000, representing an effective rate of $40,000/$500,000 = 8%. Since the Swiss franc appreciated substantially in our example, the effective financing rate was very high. If the U.S. firm had anticipated the Swiss franc's substantial appreciation, it would not have borrowed francs.

A negative effective financing rate implies the U.S. firm actually paid fewer dollars in total loan repayment than the amount of dollars borrowed. Such a result can occur if the Swiss franc depreciates substantially over the life of the loan. This does not imply a loan will be basically

The effective financing rate (called rf) is derived as follows:

$$rf = (1 + if)(1 + ef) - 1$$

where

rf	=	effective financing rate
if	=	financing rate of the foreign currency
ef	=	reflects the percentage change in the Swiss franc against the US dollar
rf	=	(1 + .08) (1 + .20) – 1
	=	.296 or 29.6%

Fig. 8-11 Effective Borrowing Rate of Foreign Currency

"free" anytime the currency borrowed depreciates over the life of the loan. Yet depreciation of any amount will cause the effective financing rate to be less than the quoted interest rate.

Consider a second example for the U.S. firm. Based on a quoted rate of 8% for the Swiss franc, and depreciation in the franc from \$.50 (on the day funds were borrowed) to \$.45 (on the day of loan repayment), what is the effective rate of a one year loan from the viewpoint of the U.S. firm? Figure 8-12 illustrates the impact.

First compute the percentage change in the Swiss franc's value:

$$C = \frac{A - B}{B}$$

where

C	=	Percentage change in the currency
A	=	Currency rate on day of loan repayment
B	=	Currency rate on day loan was granted

$$C = \frac{\$.45 - \$.50}{\$.50}$$

$$= \frac{-\$.05}{\$.50} = -.10\%$$

Next, the quoted interest rate (if) of 8%, and the percentage change in the Swiss franc (ef) of –10% can be inserted into the formula for calculating the effective financing rate (rf) by above formula in Figure 8–11.

$$rf = [(1 + .08)(1 - 10)] - 1$$

$$= [(1.08)(-.9)] - 1$$

$$= -.028 \text{ or } -2.8\%$$

Fig. 8-12 Effective Borrowing Rate of Currency When Rate of Exchange Declines

MANAGEMENT OF INVENTORIES

Inventory management determines the amount of inventory carried by the firm and, therefore, the amount of financing needed. The ability to use inventory as loan collateral is directly related to the type of inventory and how it is managed. The inventory carried by the firm is tied into production and sales activities, which are responsible in turn for the majority of the firm's cash flows. How inventory is managed will affect the length of the firm's cash flow cycle and the estimation of cash flows. This in turn has an impact on both the types of cash flow forecasting that the firm can employ and their accuracy. The timing and uncertainty of cash flows also affects the firm's liquidity needs and, potentially, how that liquidity is maintained.

Inventories provide some benefits by uncoupling elements of the purchasing, production and shipment processes. A primary benefit of holding sufficient inventory is the reduction of stockout costs, that is, reducing the lost sales and margins as well as the number of lost customers due to stockouts. In a more positive sense, inventory may provide a competitive advantage over other firms and result in increased customer goodwill and higher service levels, and therefore increased sales and profits.

In the petroleum industry crude inventories are generally under the management of the crude oil production and delivery department. The product inventory is usually managed by the operations product distribution and supply departments. Electronic Data Interchange (EDI) facilities have made significant improvement in matching the types of crude purchased by the refinery group and their processes with the requirements of the marketing department. The total oil and gas inventory includes capacities of the petroleum distribution system. In this system are crude in tankers, inventory at marine terminals, lease tankage, pipelines, terminals, refineries, product pipelines and terminals, product marine terminals and customer inventories.

Inventories are essential to the operation of most businesses, and the typical firm has 20% of its assets tied up in inventories. Inventories are commonly classified into the following categories: (1) raw materials such as crude, (2) work-in-process such as unfinished crude inventories at refineries, and (3) finished products such as jet fuel, gasoline, and natural gas, etc. Inventories, like accounts receivable, are greatly influenced by the level of, and changes in, sales volume. Since inventories are acquired before sales can take place, an accurate sales forecast is critical to effective inventory management. As inventory size increases, some costs (*ordering costs*) decrease while other costs (*carrying costs*) increase. These two costs, when added together, comprise the total cost curve for inventory. The minimum point on the total cost curve designates the optimal inventory position.

Supplier-managed inventory (SMI) is a technique oil companies are using today for material and supplies. Oil companies negotiate arrangements whereby they allocate storage space to the supplier. When the inventory items are used, the company pays the supplier for the goods. This shifts the investment in inventory from the consumer to the supplier. The objective of SMI is to reduce production and logistics costs and to share the savings with the customer.

The economic ordering quantity (EOQ) model is a common approach for determining the optimal level of inventory (Fig. 8–13). EOQ is determined as follows:

- Ordering costs that decline with larger inventories are identified, such as costs of placing orders, sales lost because of shortages, cost of production interruptions, etc.
- Carrying costs which tend to increase with larger inventories are identified, such as warehousing costs, insurance, obsolescence, interest on inventory investment.

The EOQ model is formulated as follows:

$$EOQ = \sqrt{\frac{2FS}{CP}}$$

EOQ	=	Economic Order Quantity
F	=	Fixed costs of placing and receiving an order
S	=	Annual sales in units
C	=	Carrying cost expressed as a percentage of inventory value
P	=	Purchase price per unit of inventory.

Fig. 8–13 Economic Order Quantity

The assumptions of the EOQ model do not actually hold, hence safety stocks of inventory are needed to offset changes in the rate of sales and production and to allow for delays in shipping. Since significant geographical dispersion exists in the oil and gas industry, material warehouses are used to service the repairs and maintenance requirements of assigned areas. The warehouses are maintained to ensure availability of unique parts and equipment. It is important that the inventory values at these warehouses be under continuous observation to equalize the costs of maintaining the inventories versus the cost of a shut down or operational delay.

TECHNOLOGIES AND TECHNIQUES AFFECTING WORKING CAPITAL

The advent of EDI has had the single largest impact on working capital management. EDI has ushered in the information revolution much the same way the assembly line launched the industrial revolution a century ago. EDI enables business managers to manage information about transactions and goods rather than to micro-manage the transactions and goods themselves. The materials management discipline no longer is confined to the warehouse or distribution center. It now stretches upstream to include procurement and downstream to encompass distribution and the sales function.

EDI: Working Capital Rescue

Today businesses can use EDI to control the entire purchase-sale timeline. EDI is the computer-to-computer exchange of information using standardized, machine-processable data formats. An EDI transaction set is equivalent to a paper document such as a purchase order, invoice, or bill of lading. EDI originated in the freight transportation business and quickly spread to manufacturing with the development of the EDI purchase order, invoice, and payment order/remittance advice. Today nearly 300 different EDI transaction sets have been developed and used in scores of industries.

Several factors have contributed to the popularity of EDI, including the following:

- Quicker response and shorter cycle times with EDI enable companies to acquire products and replenish their supplies faster.
- Fewer data errors result in EDI users sharply reducing their volume of manual data entries.
- Lower product acquisition costs as EDI users eliminate paper-based processes and integrate electronic data with operating and other application systems.

Several EDI transaction sets enable suppliers and their customers to exchange the information necessary to enable the supplier to manage the inventory. The American Petroleum Institute and other industry groups have spearheaded the EDI project in the petroleum industry. The EDI data transaction sets already in use include:

- Geophysical, geological and drilling activities
- Production and royalty data
- Crude sales and exchanges
- Rail, truck, ocean, and pipeline transportation methods
- Gas plant flow information
- Distribution, demand and supply data
- Regulatory, leasing, and environmental data sets
- Joint interest functions
- Material purchases and warehouse inventories
- Electronic transfer of funds, and
- Refineries and downstream marketing activities

EDI in Receivables and Payables Management

The relative availability of cash usually is the dominant determining factor of the debtor's ability to pay. However, EDI facilitates payments and, therefore, has applications for the payment and corresponding collection functions. Collection and disbursement application systems typically are computerized, but most buyers and sellers use paper (checks and paper remittance advice) to communicate. However, EDI provides a lower cost and a faster and more accurate bridge between the seller's billing and receivables system and the buyer's accounts payable system.

How EDI Helps

What does this have to do with working capital? EDI reduces the transaction cost of acquiring and selling goods for both the buyer and supplier. In addition, the buyer can take purchase discounts that it otherwise might have lost in the slow-moving paper environment. Also, purchasing personnel control both pricing and payment terms. This eliminates the need to have an accounts payable clerk looking for discrepancies.

EDI enables suppliers to apply cash sooner because of the automated processing. This releases the customer's credit line sooner, enabling greater sales at the expense of competitors at times when the buyer's credit line is at its limit. EDI also benefits the customer as well by freeing the buyer's credit limit sooner, thereby enabling the customer to be more flexible in both purchasing and the timing of receipt of goods.

SUMMARY

Working capital, sometimes called gross working capital, simply refers to current assets. Net working capital is current assets minus current liabilities. Working capital policy refers to basic policy decisions regarding target levels for each category of current assets and to the financing of these assets. Working capital management involves the day-to-day administration, within policy guidelines, of current assets and current liabilities.

Working capital management begins in purchasing, on the shop floor, in the interface with suppliers and customers, and in shipping far outside the walls of the accounting department. Working capital is nominally an accounting concept, but it must be funded by real sources of long-term and permanent capital. If the corporate objective is to maximize stock-holder value over the long run, a company must limit its investment in working capital, while still maintaining adequate liquidity for normal operations.

The financial manager should be as concerned about short-term financing as long-term financing. Cash flow is critical to maintaining credit rating and relationships with suppliers. If current cash flow is not managed properly, the company could find itself forced into bankruptcy.

In addition, creditors review working capital and current ratios to evaluate credit-worthiness of the company. Extending credit is often decided upon the net working capital position of the company.

Large inventories have caused financial problems within companies as significant capital is tied up in inventory which does not have earning power and the expense of maintaining inventory eats away at profit. New concepts of just-in-time (JIT) inventories have an objective of zero inventory by creating partnerships with suppliers to provide materials only as needed. Obviously this will relieve the financial manager from the headache of minimizing capital tied up in inventories.

Accounts receivable financing involves either the pledging of receivables or the selling of receivables (factoring). Pledging of receivables is characterized by the fact that the lender not only has a claim against the receivables, but also has recourse to the borrower (seller). Factoring involves the purchase of accounts receivables by the lender without recourse to the borrower (seller).

A large volume of credit is secured by business inventories. If a firm is a relatively good credit risk, the mere existence of the inventory may be a sufficient basis for receiving an unsecured loan.

Capital Structure and Cost of Capital

9

The petroleum industry's worldwide upstream spending in 1995 was $59.3 billion. About 68% or more of this amount went toward leasing, exploration, drilling, and production activities. The magnitude of these figures may be difficult for the average person to grasp. But it is plain to see that the industry is *capital intensive*. In simple language, this means the oil business requires an inordinate amount of capital for operations, particularly to finance exploration, drilling, and production. Relative to production, these figures also include secondary and tertiary expenditures.

Where, one may ask, do these huge sums originate? Where does the *risk capital* come from to drill, complete, and equip or plug thousands of oil and gas wells that are drilled each year; to upgrade refineries to produce environmentally acceptable product; or to build ocean tankers, pipelines, barges, trucks or rail cars to transport crude, gas, and products to their destination? The capital for this sizable undertaking comes from several sources. Some capital is generated within the industry from operations. Corporations and large independents customarily reinvest a percentage of profits for the leasing, exploration, and drilling activities to maintain growth and profitability within the company. To supplement these amounts taken from profits, companies may sell stocks, issue bonds, or borrow against current production.

This chapter is concerned with whether the way in which investment proposals are financed matters; and, if it does matter, what is the optimal capital structure? Consideration will also be given to the calculation of the cost of capital used as a hurdle rate to ration funds for capital projects.

If one mix of securities rather than another is used to finance projects, is the market price of the company's stock affected? If the firm can affect the market price of its stock by its financing decision, it will want to undertake a financing policy that will maximize market price. This prompts the question of capital structure in terms of the proportion of debt to equity. The specific type of security being issued is also an important consideration.

"Capital Structures"

CAPITAL STRUCTURE DECISIONS CHECKLIST

In determining the sources of funds and their impact upon the firm, critical issues must be given consideration. The following checklist of items should be considered in determining the composition of the corporate capital structure.

Sales Stability

If sales are relatively stable, a firm can safely take on more debt and incur higher fixed charges than can a company with unstable sales. Utility companies, because of their stable demand, have been able to undertake more debt financing than industrial firms. The oil industry can project volumes and consumer prices, generally with a degree of accuracy. However, crude prices are affected by unpredictable and uncontrollable forces.

Asset Structure

Firms with assets suitable as security for loans tend to use debt rather heavily. Thus, real estate companies are often highly leveraged, while companies involved in technological research employ less debt.

Operating Leverage

All things being equal, a firm with less operating leverage is better able to employ financial leverage. Operating leverage occurs any time a firm has fixed costs that exceed variable costs. The large sums of money required for investment by the downstream functions of the oil and gas industry can generate unused capacity that negatively affects operating profits.

Growth Rate

Typically, faster growing firms must rely more heavily on external capital. Further, the flotation costs involved in selling common stock exceed those incurred when acquiring debt. Thus, rapidly growing firms tend to use somewhat more debt than do maturing or slower-growth companies. The oil industry is in the maturing stage.

Profitability

Firms with very high rates of return on investment often use relatively little debt. The practical reason is that very profitable firms simply do not need much debt financing. Their high rates of return enable them to do most of their financing with retained earnings and still have adequate funds to pay to the owners.

Taxes

Interest payments on debt are a deductible expense, while dividends are not deductible. Hence, the higher a firm's corporate tax rate, the greater the advantage of using debt. Borrowing is generally considered to cost less than equity funds because of the tax benefits.

Control

The effect that debt or stock financing might have on a management's control position may influence its capital structure decision. If management has voting control (over 50% of the stock) but is not in a position to buy any more stock, debt may be the choice for new financing. On the other hand, a management group that is not concerned about voting control may decide to use equity rather than debt. If the firm's financial situation is too weak, the use of debt might subject the firm to serious risk of default. If the firm goes into default, some financial managers may lose their jobs.

However, if too little debt is used, management runs the risk of a takeover attempt; some other company or management group may plan to boost earnings and stock prices by using financial leverage. In general, control considerations do not necessarily suggest the use of debt or equity, but if management is at all insecure, the effects of capital structure on control are certainly taken into account.

Lender and Rating Agency Attitudes

Regardless of managers' analyses of the proper leverage for their firms, there is no question but that lenders' and rating agencies' attitudes are frequently important determinants of financial structure. In the majority of cases, the corporation discusses its financial structure with lenders and rating agencies and gives much weight to their advice. But when management is so confident of the future that it seeks to use leverage beyond the norms for its industry, lenders may be unwilling to accept such debt increases, or may do so only at a high price.

Market Conditions

Conditions in the stock and bond markets undergo both long- and short-run changes that can have an important bearing on a firm's optimal capital structure. For example, during the credit crunch in the first part of the 1980s, there was simply no market at any "reasonable" interest rate for new long-term bonds rated below an "A" rating. Low-rated companies with a "B" or "C" rating that needed capital were forced to go to the stock market or to the short-term debt market.

Actions such as this could represent either permanent changes in target capital structures or temporary departures from stable targets. The important point is that stock and bond market conditions do influence the type of securities used for a given financing.

The Firm's Internal Conditions

A firm's own internal conditions have a bearing on its target capital structure. For example, suppose a firm has just successfully completed a research and development (R&D) program and projects higher earnings in the immediate future. However, the new earnings are not yet anticipated by investors, hence are not reflected in the price of the stock. This company would not want to issue stock. It would prefer to finance with debt until the higher earnings materialize and are reflected in the stock price--at which time it might want to sell an issue of common stock, retire the debt, and return to its target capital structure.

COMPANY'S GROWTH, DEVELOPMENT AND FINANCIAL REQUIREMENTS

As a company grows, it usually requires additional funds to finance working capital, plants and equipment, as well as for other purposes. The company could issue additional shares of stock, but this might dilute earnings per share for a time or perhaps raise questions of control. Another alternative is to borrow long-term funds. As a result, the remaining source of long-term capital (excluding some assets sales, etc.) is the growth in retained earnings. Such a method is typically a slow way to gain additional capital. The rate of growth of equity is important for establishing target rates of return on equity, selecting sources of capital and monitoring dividend policy.

The annual growth in shareholders' equity from internal sources may be defined as the rate of return earned on such equity multiplied by the percentage of the earnings retained. It may be represented by the formula in Figure 9–1.

Growth in Shareholders' Equity (G) = Return on Equity (R) x (1 – Payout Ratio "P")

As an example, if a company can earn about 23% each year on its equity, and the payout ratio or the percent of earnings that go to dividends is 40%, then shareholders' equity will grow at 14% per year, calculated as follows:

$$
\begin{aligned}
G &= R(1 - P) \\
&= 0.23\,(1 - 0.40) \\
&= 0.23\,(0.60) \\
&= 0.14 \\
&= 14\%
\end{aligned}
$$

Fig. 9–1 Definition of Internal Growth

If management thinks the company can grow in sales and earnings at about 30% per year; additional funds will be needed at about this same rate. And if the dividend payout is to remain at 40%, then management will require some outside capital for the growth potential to be realized.

RETURN ON EQUITY AS RELATED TO GROWTH IN EARNINGS PER SHARE (EPS)

Another facet of the shareholders' equity (outstanding stock) is the relationship of the return on equity to the rate of annual increase in earnings per share. This connection is often not understood even by some financial executives. Basically, the rate of return on shareholders' equity, when adjusted for the payout ratio, produces the rate of growth per year in EPS. It is expressed by the formula in Figure 9–2.

Growth per Year in EPS (G) = Return on Equity (R) x [1 – Retention Ratio (P)]

Thus, assuming a constant return (R) on equity of 20%, a constant dividend payout ratio (P) of 25%, a retention ratio (1 – payout ratio) of 75%, the EPS growth rate is calculated by means of the same formula as for the growth of shareholders' equity:

$$\begin{aligned} G &= R(1 - P) \\ &= 0.20(1 - 0.25) \\ &= 0.20(0.75) \\ &= 0.15 \\ &= 15\% \end{aligned}$$

Fig. 9–2 Relationship of Return on Equity to Earnings per Share (EPS)

Growth in Earnings per Share

Prudent financial planning will consider the impact of decisions on earnings per share (EPS). Management is generally concerned with the growth in EPS, since one of its tasks is to enhance shareholder value. Continual increases in EPS will raise shareholder value through its recognition in a higher price/earnings (P/E) ratio and usually a rising dividend payment. Moreover, the growth in EPS is one of the measures of management as viewed by the financial community, including financial analysts.

Given the importance of EPS, financial officers should bear in mind that the earnings per share will increase as a result of any one of the following actions:

- Plow-back of some share of earnings, even as long as the rate of return on equity remains just constant (however, a growth in EPS does not necessarily mean that management is achieving a higher rate of return on equity).
- Actual increase in the rate of return earned on shareholder's equity.
- Repurchase of common shares as long as the rate of return on equity does not decrease.
- Use of prudent borrowing or financial leverage.
- Acquisition of a company whose stock is selling at a lower P/E is better than acquiring a company with a higher P/E.

- Sales of shares of common stock above the book value of existing shares, assuming the return on equity (ROE) is maintained.

BUSINESS RISK

Business risk is defined as the uncertainty inherent in projections of future operating income, or earnings before interest and taxes (EBIT). The single most important determinant of a firm's capital structure is EBIT. Fluctuations in business risk are caused by many factors—inflation and recessions in the national economy, successful new products introduced both by the company and its competitors, labor strikes, price controls, new oil discoveries, and so on. Further, there is always the possibility that a long-term disaster might strike, permanently depressing the company's earning potential. For example, a competitor could introduce a new product that might permanently lower company earnings, or a company could experience a succession of dry holes.

Business risk depends on a number of factors, the most important of which include:

- *Demand variability*. The more stable the demand for a firm's products, other factors held constant, the lower its business risk.
- *Sales price variability*. Firms whose products are sold in highly volatile markets are exposed to more business risk than similar firms whose output prices are more stable.
- *Input price variability*. Firms whose input prices are highly uncertain are exposed to a high degree of business risk. The high uncertainty of crude oil prices constantly haunts the financial planner.
- *Output prices*. Some firms are better able than others to raise their own output prices when input costs rise. The greater the ability to adjust output prices, the lower the degree of business risk. This factor has become increasingly important during periods of inflation.
- *Fixed costs*. If a high percentage of a firm's costs are fixed, hence do not decline when demand falls off, then it is exposed to a relatively high degree of business risk. A refinery high fixed investment costs and faces risk if demand decreases or environmental laws require new processes.

Each of these factors is determined partly by the firm's industry characteristics, but each of them is also controllable to some extent. For example, most firms can, through their marketing policies, take actions to stabilize both unit sales and sales prices. However, this stabilization may require spending a great deal on advertising, and/or price concessions in order to get customers to commit to purchase fixed quantities at fixed prices in the future. Similarly, firms may reduce the volatility of future input costs by negotiating long-term labor and materials supply contracts, but they may have to agree to pay prices above the current spot price level to obtain these contracts.

Business risk depends in part on the extent to which a firm builds fixed costs into its operations. If fixed costs are high, even a small decline in sales can lead to a large decline in EBIT. So all things being equal and the higher a firm's fixed costs, the greater its business risk. Higher fixed costs are generally associated with more highly automated, capital-intensive firms and industries. Also, businesses that employ highly skilled workers who must be retained and paid even during recessions have relatively high fixed costs.

LEVERAGE

There are two dimensions of risk. Operational risk is the extent to which a firm builds fixed costs into its operation. If fixed costs are high, even a small decline in sales can lead to a large decline in operating earnings before interest and tax (EBIT). So when other things are held constant, the higher a firm's fixed costs, the greater its business risk. Financial risk is the additional risk placed on the common stockholders as a result of the firm's decision to use debt.

Operational Leverage

A company with a high percentage of fixed costs is more risky than a firm in the same industry that relies more on variable costs for operations. The degree of operating leverage (DOL) is the percentage change in net operating income associated with a given percentage change in sales (Fig. 9–3).

$$\text{Degree of Operating Leverage} = \frac{\%\ \text{change in net operating income}}{\%\ \text{change in sales}}$$

Operating income increases 20% with a 10% increase in sales

$$\text{DOL} = \frac{20\%}{10\%}$$

$$= 2.0$$

Fig. 9–3 Calculation of Degree of Operating Leverage

Financial Leverage

Financial leverage is the relative amount of the fixed cost of capital, principally debt, in a firm's capital structure. Leverage, by definition creates financial risk, which relates directly to the question of the cost of capital. The more leverage, the higher the financial risk, the higher the cost of debt capital:

- *Earnings per share* (EPS) will ordinarily be higher if debt is used to raise capital instead of equity, provided that the firm is not overleveraged. The reason is that the cost of debt ordinarily is lower than the cost of equity because interest is tax deductible. However, the prospect of higher EPS is accompanied by greater risk to the firm resulting from required interest costs, creditors' liens on the firm's assets, and the possibility of a proportionately lower EPS if sales volume fails to meet projections.
- *Degree of financial leverage* (DFL) is the percentage change in earnings available to common shareholders associated with a given percentage change in net operating income. If the return on assets exceeds the cost of debt, additional leverage is favorable (Fig. 9–4).

$$\text{Degree of Financial Leverage} = \frac{\%\ \text{change in net income}}{\%\ \text{change in net operating income}}$$

1. Net income means earnings after taxes available to common shareholders
2. Operating income equals earnings before interest and taxes (EBIT)
3. The greater the DFL, the riskier the firm

Fig. 9–4 Calculation of the Degree of Financial Leverage (DFL)

Whereas operating leverage refers to the use of fixed operating costs, financial leverage refers to the use of fixed charge securities—debt and preferred stock. Conceptually, the firm has a certain amount of risk inherent in its operations. This is its business risk, which is defined as the uncertainty inherent in projections of future EBIT. If a firm uses debt or financial leverage, this concentrates its business risk on the stockholders.

For example, suppose 10 people decide to form a corporation to manufacture solar fuel cells. There is a certain amount of business risk in the operation. If the firm is capitalized only with common equity, and if each person buys 10% of the stock, then the investors all share the business risk.

However, suppose the firm is capitalized with 50% debt and 50% equity, with five of the investors putting up their capital as debt and the other five putting up their money as equity. In this case, the investors who put up the equity will have to bear most of the business risk, so the common stock will be twice as risky as it would have been had the firm been financed only with equity. Thus, the use of debt concentrates the firm's business risk on its stockholders.

Beta or Market Risk

Beta (or market) risk is the risk that is relevant to most stockholders. Therefore, a firm can be risky in a business risk sense (that is, there can be great uncertainty about its future EBIT), but if its returns are not perfectly correlated with those of other firms, then the stock may still be regarded as being not very risky by diversified investors. In other words, a part of any firm's business risk is company-specific, or *unsystematic risk*.

Since company-specific risk can be eliminated by diversification, an investor who holds a stock in isolation must bear his or her share of the entire business risk of the firm. By definition, however, well-diversified stockholders render "irrelevant" the company-specific portion of the firm's business risk. Still, the high correlation among returns on different firms generally indicates that most increases in business risk result in higher systematic or beta risk. This has an adverse effect even on well-diversified investors.

Even though debt has a prior claim on the firm's assets and income, under extremely bad business conditions, debt holders can still suffer losses. Therefore, debt holders cannot be totally protected against business risk. To illustrate this concept, consider a firm with an expected EBIT of $4 million and assets of $20 million. We first divide EBIT by the dollar amount of assets required to produce the EBIT, obtaining the rate of return on assets (ROA), which in this case is:

$$ROA = \$4/\$20 = 0.20 = 20\%$$

If the company had no debt, then:

- Its assets would be equal to its equity
- Its return on equity (ROE) would be equal to its return on assets (ROA), and
- Its equity would be exactly as risky as the assets

Now suppose that the firm decides to change its capital structure by issuing $10 million of debt which carries a 15% interest rate and then uses these funds to retire equity. Its expected return on equity (which would now be only $10 million) would rise from 20% to 25% (Fig. 9–5).

Expected EBIT (unchanged)	\$ 4,000,000
Less interest (15% on \$10 million of debt)	–1,500,000
Income available to common stock (zero taxes)	\$ 2,500,000

Expected ROE = \$2,500,000/\$10,000,000 = 25%

Fig. 9–5 Effects of Debt on ROE

Thus, the use of debt would "leverage" expected ROE up from 20% to 25%. However, financial leverage also increases risk to the equity investors. For example, suppose EBIT actually turned out to be \$2 million rather than the expected \$4 million. If the firm had not used debt, the ROE would have declined from 20% to 10%; but with \$10 million debt, ROE would have fallen from 25% to only 5%.

Interrelationship between Financial and Operating Leverage

The degree of operating leverage (DOL) is defined as the percentage change in operating profits associated with a given percentage change in sales volume. The calculation is illustrated in Figure 9–6.

$$\text{Degree of operating leverage at Point Q} = \frac{Q(P - V)}{Q(P - V) - F}$$

or

$$DOL = \frac{S - VC}{S - VC - TF}$$

where

Q	=	units of output = 1,000 barrels
P	=	average sales price per unit of output = \$20
S	=	Sales in total dollar = \$20,000
V	=	variable cost per unit = \$8.00
VC	=	Total variable costs = \$8,000
F	=	fixed operating costs = \$10.00
TF	=	Total Fixed Operating Cost = \$10,000

$$DOL = \frac{\$20{,}000 - \$8{,}000}{\$20{,}000 - \$8{,}000 - \$10{,}000} = \frac{\$12{,}000}{\$\ 2{,}000} = 6$$

Fig. 9–6 Calculation of Operating Leverage

A low DOL indicates that there are a lot of fixed costs incurred in the sale of one barrel of crude oil. Variable costs are a function of throughput; however, fixed costs are not affected by throughput in the short-term until capacity is totally utilized.

Total Leverage

The degree of total leverage (DTL) combines the DFL and the DOL. It equals the degree of financial leverage times the degree of operating leverage. Thus, it equals the percentage change in net income associated with a given percentage change in sales as illustrated in Figure 9–7.

$$\text{DTL} = \text{DFL} \times \text{DOL}$$

$$= \frac{\%\text{ change in net income}}{\%\text{ change in net operating income}} \times \frac{\%\text{ change in net operating income}}{\%\text{ change in sales}}$$

$$= \frac{\%\text{ change in net income}}{\%\text{ change in sales}}$$

If net income increases 15% with a 5% increase in sales

$$\text{DTL} = \frac{15\%}{5\%} = 3.0$$

Fig. 9–7 Calculation of Degree of Total Leverage

Firms with a high degree of operating leverage do not usually employ a high degree of financial leverage and vice versa. One of the most important considerations in the use of financial leverage is operating leverage.

Example: A refinery has a highly automated production process. Because of automation, the degree of operating leverage is 2. If the firm wants a degree of total leverage not exceeding 3, it must restrict its use of debt so that the degree of financial leverage is not more than 1.5. If the firm had committed to a production process that was less automated and had a lower DOL, more debt could be employed and the refinery could have a higher degree of financial leverage.

BUSINESS FORMS AND FINANCING

The legal form of a business will affect financing. The first step in organizing a business is to determine the economic unit for which financing, legal, accounting and tax objectives can be achieved. A business entity can be identified as *sole proprietorship* owned by one individual; a *partnership* owned by two or more individuals; or a *corporation* organized under state or federal laws as a separate legal entity. The ownership of a corporation is divided into shares of stock. The sole proprietorship is the most common business form. However, corporations receive over 90% of the total dollars of business receipts.

Forms of Business Organization

The three basic forms of business organizations are: sole proprietorships, partnerships, and corporations. Each form is recognized as an economic unit separate from its owners, although legally only the corporation is considered separate from its owners. Other legal differences among the three forms are summarized in Table 9–1 and discussed briefly below:

	Sole Proprietorship	**Partnership**	**Corporation**
1.Legal Status	Not a separate legal entity	Not a separate legal entity	Separate legal entity
2.Risk of Ownership	Owner's personal resources at stake	Partner's resources at stake	Limited to investment in corporation
3.Transferability of Ownership	Sale by owner establishes new company	Changes in any partner's % of interest requires new partnership	Transferable by sale of stock
4.Accounting Treatment	Separate economic unit	Separate economic unit	Separate economic unit

Table 9–1 Comparative Features of the Forms of Business Organization

Limited Partnerships

A great deal of equity and debt capital is raised each year by issuance of securities registered with the Securities and Exchange Commission (SEC). In fact, in many industries, virtually all of the equity capital raised by established companies is raised in this manner. In a few industries, however, a significant amount of capital is raised from private investors through the use of securities not registered with the SEC. The industries in which this is most prevalent are real estate and oil. In these industries, the primary manner in which private funds are raised is through limited partnerships.

In private financing, a firm in essence approaches one or more private investors in an attempt to convince them to invest funds in a partnership created to finance one or more aspects of the sponsoring firm's activities. A limited partnership might be formed, for example, to finance the drilling of one or more wells, or to finance a pipeline project, with the partnership retaining a specific percentage interest in the cash flows that hopefully will be generated by the projects.

Master Limited Partnership

In recent years significant capital has been raised by master limited partnerships. Unlike the limited partnerships, master limited partnerships must have their offerings and units approved by the SEC and are then sold through one of the public stock exchanges. The general partners are the operators, and the limited partners are investors with limited participation in the operations.

Joint Ventures

Still another form of raising money in the oil industry—one that is far less common in most other industries—is the joint venture. Normally, the process involves putting together one or more deals in which one of the parties contributes either assets or expertise, while the other party or parties contribute money. Also common is the arrangement whereby a small group of firms contributes funds and appoints one of the firms as operator of the specific project. Joint venture arrangements might involve only large firms, only small firms, or some combination of the two.

If the joint venture is an *undivided interest* project, the assets, liabilities, expenses and revenues are recorded on the joint interest owner's books in proportion to their ownership ratios. The operator usually makes an advance cash call to the joint interest owners for operating costs based on projections for the future period costs. An account is provided for actual spending versus budget. The over/under cash balance is carried to the future cash call. The undivided interest owners are responsible for raising the capital required by the joint venture, either from operations, debt or capital. Many interest owners may be involved in an undivided interest project.

It is not unusual for the joint venture to take the business form of a corporation with the participating owners as stockholders. Either one of the affiliate owners may operate the joint venture under the provisions of an *operating agreement* and bill an operating fee to the other joint venture owners or the joint venture may operate independently of the stockholder companies. Debt capital is carried on the books of the joint venture and often guaranteed with a *throughput and deficiency agreement* by the stockholders. In this way the stockholder companies are not required to inject capital into the joint venture.

If the stockholders do not have control of the joint venture i.e., owning more than 50%, the joint venture is accounted for on the stockholder's book using the *equity method*. The debt is not consolidated on the stockholder's record. An objective of most public companies is to maintain a low debt/equity ratio that reduces risk and protects the bond ratings of the corporation. The equity method does not require the recording of debt on the stockholder company's financial records. However, a disclosure is required in the accompanying notes to the financial statements reporting the debt exposure by equity joint ventures.

Firms in the oil industry often employ peculiar forms of financing (or financial vehicles). Large-scale project financing, the extensive use of limited and master limited partnerships, and the frequent formation of both financial and operational joint ventures are common to only a few industries. Some of the problems faced by oil firms in the financial planning and control area are ones that are either common to only a few industries or substantially more pronounced in these industries.

Two common problems faced by the financial planners in oil firms when joint ventures are concerned include (1) the impact of cost allocations by the operator to joint ventures, and (2) the cost of capital used in analyzing capital expenditures.

1. The operating company's financial planners need to consider the impact of reduction of costs on the operator's records because of charge outs to joint ventures. Conversely, the planners of the nonoperating company need to consider the impact of increased costs charged them from the operating company.
2. In the capital expenditure analysis, the financial analyst of the operating company will have knowledge of only their cost of capital, whereas capital decisions may be entirely different if the nonoperating cost of capital was used.

The oil and gas industry has developed creative strategies and techniques in mitigating high levels of risk and increasing costs. Costs of borrowed capital are escalated by high-risk levels in the normal channels of borrowing, therefore unique business arrangements and forms have been developed to acquire financing through different channels.

Private Venture Capital

The largest source of noncorporate funds for drilling and production activities is the deep reservoir of private capital. Individuals with money to lose and hopes of striking it rich put millions of dollars into the oil business by investing in large drilling funds and smaller private funds. Others subscribe to the programs of independent drilling and production companies and invest in one- or two-well programs with independent operators.

Independent Operators

For the sake of clarity, a distinction should be made between the independent operator and the independent drilling and production company. The independent operator is usually a small, three-to-five person organization or partnership engaged in a one- or two-well drilling program at any one time. The independent operator is strictly an oil producer; this organization has no downstream interests. As a rule, it hires a drilling contractor to put down its wells.

By contrast, the independent drilling and production company may have hundreds of employees. It also may have a number of drilling rigs used for its own drilling programs as well as for contract work. Larger independents usually have some downstream operation, such as a gas or oil gathering system or a gas stripping plant.

In many important ways, both the small operator and the independent company make their presence felt in the oil business. For example, of the 76,000 wells drilled in 1982, approximately 85% were drilled by independents. And to show who the risk-takers are, of the 17,000 wildcat wells drilled that year, only about 2,000 were drilled by the 20 largest companies. The other 15,000 were drilled by the independents.

As might be expected, competition is intense between large and small companies to obtain favorable prospects. Despite this rivalry, there is a long-established cooperation between these companies that serves them both. Although independents drill most of the onshore wells, large corporations contribute substantially to independent drilling programs in farmout acreage and dry-hole and bottom-hole monies. As a result, thousands of acres of land leased by major corporations are tested and evaluated. The spirit of adventure is still very much alive in this business whose cachet is risk and expectations.

Oil Field Deals

The exploration and production segment has developed creative ways of mitigating risk and expense through special joint interest ventures. These procedures, used by the working interest owner, assign rights to other parties who will share in financing and costs. First, we will discuss the ownership of oil and gas rights, then how working interest is assigned to other interested parties to provide the risk funds.

Oil and Gas Rights

Exploration and production companies normally get the right to explore for and extract oil and gas by leasing that right from the owner of the mineral rights on a piece of property. The mineral rights may or may not be owned by the same person who owns the surface rights. The owner of the mineral interest may grant a leasehold interest. In that instance the mineral interest owner becomes the *lessor*, and the owner of the leasehold interest is the *lessee*. The lessor usually retains a royalty interest (usually one-eighth interest) as compensation for allowing someone else to extract the minerals in place.

A *leasehold interest* is conveyed by the mineral interest owner (lessor) to the lessee. The leasehold interest owner, often called the *working interest owner*, has the right to explore and produce minerals. A 100% working interest owner will pay all of the costs to develop and produce the mineral and will receive all the revenue less any royalty interest retained by the mineral interest owner in the leasing transaction. Since the leasehold interest is created by a legal agreement, it is subject to any provisions contained in the agreement. Common lease provisions are: the primary term, lease bonus, royalty interest, delay rentals, shut-in gas well provisions, pooling and cessation of production clauses.

Primary Term

Leases are usually in effect for a period of 1 to 10 years (called the *primary term*) and as long thereafter as hydrocarbons are produced. When negotiating for a lease, the landowner (mineral interest owner) usually tries to get as short a primary term as possible. The lessee wants as long a term as possible since this gives him more time to decide where to drill.

Lease Bonus

The lessee pays a "sign-up" bonus to the lessor in order to induce him to sign. The amount ranges from $1 per acre for goat pasture to over $3,000 per acre for very good property.

Royalty Interest

In addition to the bonus, the lessee agrees to pay the lessor a certain percentage of the money received from the sale of hydrocarbons. The lessor does not have to pay any of the costs of drilling and equipping the wells. The lessor also does not pay any of the costs of treating the hydrocarbons except under certain conditions (typically compression, treating of, or transportation of gas). The lessor does have to pay his own severance and ad valorem taxes.

In the past, the amount of the royalty was almost always one-eighth of the gross proceeds. Now it is not unusual to see royalties of three-sixteenths or one-fourth. In certain areas that are known to contain hydrocarbons, the federal government has a sliding scale royalty.

Delay Rental

Unless operations are commenced on or before one year from the effective date of the lease, the lease terminates unless the lessee pays a delay rental. This delay rental allows the lessee to defer the commencement of operations for one year. At the end of that year, the lessor can again pay the delay rental and hold the lease. This process continues until the primary term expires or production is established. Typical delay rentals are $1/acre although the lessor can put pressure on the lessee to drill by providing for increasing delay rentals.

Shut-In Well Clause

Many leases allow the lessee to hold the lease beyond the primary term by paying an amount equal to the annual delay rental if there is a completed gas well on the lease (or on acreage pooled with the lease) which is shut in. This provision allows the lessee time to develop enough gas reserves to justify a pipeline. In practice it can be used to hold a lease indefinitely at little additional cost if a marginal gas well is drilled. The lessor can thwart this by having large delay rentals or by limiting the amount of time that the shut-in royalty will hold the lease. Two to five years beyond the primary term is reasonable.

Cessation of Production Clause

Leases normally provide that if a well becomes incapable of production, the lessee must commence reworking or drilling operations within 60 days or lose the lease (if the primary term has expired). This can become critical if the only well on a lease goes down for two months. The lease could be lost.

Pooling/Unitization Clause

Pooling and unitization, strictly defined, have different meanings. Pooling applies to the aggregation of small areas of land necessary to form a drilling unit (acreage needed for one well) as defined by state regulations. Unitization, however, applies to larger areas. Usually the unitized area includes the entire reservoir. Consolidation is necessary for efficient secondary recovery operations such as a water-flood. Without specific pooling or unitization language in the lease agreement, the lessee may not commit this leased area to a unit or pool his leased acreage without the consent of the royalty owner. A pooled or unitized lease will remain in effect as long as the unit remains in effect. Royalty payments are made based on the way production is allocated in the unit.

FINANCING WELL DRILLING AND DEVELOPMENT THROUGH ASSIGNMENT OF WORKING INTERESTS

The owner of the leasehold interest (or working interest owner) may convey or assign rights provided for by the lease agreement. These rights terminate when the lease expires. Overriding royalty interests, carried interests, reversionary interests, and net profits interests are some examples discussed in the following financing alternatives.

Financing Alternative #1, Overriding Royalty Interest

There are ways of acquiring drilling rights without owning a lease. This is done through a *farmout agreement*. An operator who is not ready to drill on his leased acreage assigns the lease or a portion of it to another operator who wishes to test his luck by drilling a well. As compensation, the owner of the lease usually retains a small fractional interest in the farmed-out acreage and any well to be drilled on it. The retained interest is called an overriding royalty.

An overriding royalty or override is similar to a landowner's royalty in that the holder of an override receives a portion of the production free and clear of all costs. An override is often said to be carved out of the working interest since it is created from the leasehold's share of the revenue. An override typically arises from the actions of a lease broker who obtains a lease from a landowner and then "turns it" or sells it to an oil company retaining an overriding royalty (typically 1-5% of 100%).

Farmout agreements, probably the most common type of agreement between oil companies, often provide for an override before payout convertible to a working interest after payout. The leasehold interest owner (farmor) of a block of acreage makes a deal (farms out) with another company (the farmee) whereby the farmee will drill a well at its sole risk in return for an interest in a block of acreage. Typically the farmor will retain a 5% override on the initial well until it pays out. At that time the farmor has the option to convert his override to a 25% working interest in the well. Ordinarily, all subsequent wells are paid for on a 75/25% basis whether or not the first well has paid out at the time of drilling the subsequent wells.

The amount of the override and back-in will vary, of course, 7.5% overrides and 50% back-ins were not unusual during the great turmoil (1977-1981). Sometimes, if the block of acreage is large, there will be a multiple-well commitment instead of a single-well commitment.

Financing Alternative # 2, Reversionary Interest

In the above farmout, the farmor is said to have a reversionary interest. His override reverts to a working interest at some specified point usually payout. A *back-in* is essentially the same as reversionary interest. The farmor might have a 25% back-in on the acreage.

The best way to describe a reversionary interest or back-in is " a 5% overriding royalty convertible to a 25% working interest at payout, proportionately reduced." All deals are assumed to be proportionately reduced unless it is specifically stated otherwise. If a deal is burdened with an overriding royalty of 5%, and I take one-fourth of the deal, I only pay my proportionate share of the royalty. In this case I pay 25% of 5% or 1.25%.

Financing Alternative # 3, Carried Interest

Carried interest is an interest that is not responsible for any costs and does not receive any revenue until a certain condition is met. A farmout could be expressed as a carried interest. Carried interests, however, are more likely to be a carry to the casing point of the first well or through the tanks on the first well (rather than to payout).

If a company has a 25% carry to the casing point, then the carrying party pays all of the dry-hole costs. In this case, dry-hole costs are the cost to drill, log, and test the well prior to setting the casing. At the casing point, a carried party having a 25% of carried interest has the option of putting up 25% of the completion costs to have a 25% working interest (with its associated net revenue interest) in the well and the other acreage. For a nonproducer, dry-hole costs would also include the cost to abandon the well. This is a common way for promoters to get a risk-free look at a play. If the well is a good one, they may even be able to borrow the completion money.

A one-third for one-quarter is a farmout agreement sometimes referred to as a *standard Rocky Mountain deal*. The farmee pays one-third of the costs to some predetermined point in order to receive a 25% working interest (WI). Usually, the reversion point is payout. This is the same deal as a 25% back-in after payout. Usually, there is an override before payout which converts to a 25% working interest. Notice that if all of the deal is sold, then the farmor pays nothing and gets a 25% WI after payout. This is a pretty standard deal.

Financing Alternative #4, Net Profits Interest

A net profits interest receives a portion of the net proceeds from a well after all costs have been paid. Usually the costs are cumulative. That is, the net profits interest does not receive any part of the proceeds until payout occurs.

A net profits interest appears quite similar to a carried interest or a back-in. The major difference is that a carried interest or back-in shares in the costs and receipts after specified conditions have been met (casing point, payout, etc.) while a net profits interest usually continues for the duration of production. One party continues to bear costs and the other receives a share of the proceeds after the payment of those costs.

A net profits interest will end up with the same cumulative net cash flow from a successful well as the identical percentage working interest. On an unsuccessful well, however, a working interest will lose money while a net profits interest will not share in the losses. You might think of it as a working interest without risk.

Financing Alternative #5, Production Payment

A production payment is a share of the hydrocarbons produced from the lease, free and clear of the costs of production, terminating when a given amount of hydrocarbons has been produced or when a given amount of money from the sale of that hydrocarbon has been realized. It is often used to pay down debt since the amount of money going for debt service is proportional to the producing rate.

Financing Alternative #6, Bottom-Hole Agreement

Another type of financial backing available to an independent operator is the bottom-hole agreement. This is an agreement in which an operator, planning to drill a well on his own leased acreage, secures the promise from nearby lease owners to contribute to the cost of the well. The money is paid when the well reaches the contract depth whether it is a producer or a dry hole. The contributors study the well logs, core samples, and drilling records to learn what formations lie beneath their leases so they can decide whether to drill. Oftentimes, bottom-hole agreements are used by the drilling party as collateral for obtaining a loan to help finance the well.

Financing Alternative #7, Dry-Hole Agreement

Closely related to the bottom-hole agreement is the dry-hole agreement, another example of cooperation among risk-takers. Dry-hole monies are contributions made by several interested parties (those owning land or holding leases nearby) to an operator who drills a noncommercial well or a dry hole. Before the operator rigs up to drill the well, he solicits dry-hole contributions. In return for the financial help, he agrees to furnish well logs, coring data and all drilling records to the contributors. The geological information helps determine whether to drill on adjacent acreage. If the well is dry, the operator is not obligated to repay the funds. If the well is a producer, he or she pays off the loan out of production.

JOINT OPERATING AGREEMENTS

Operating Interest

This is a term used by land and legal departments and by the IRS. The operating interest for IRS purposes is any "interest in minerals in place that is burdened with the cost of development and operation of the property." The operator of the well, on the other hand, is the entity that is responsible for the day-to-day management activities. The operator prepares cost estimates, gets approvals from nonoperators, contracts with the drilling company, supervises the drilling of the well, designs and supervises the completion of the well and does all the accounting. He is reimbursed for his costs by the nonoperators.

Nonoperating Interest

A nonoperating interest is a term that may also have dual meanings depending on its usage. A nonoperating interest in taxation refers to royalty, overriding royalty, and net profits interests. That is, any interest that is not burdened with the costs to find, develop, and produce the minerals. A nonoperator or nonoperating interest may also mean the working interest owners that are not responsible for the day-to-day operational decisions that the operator of a well must make. Working interest owners, however, are burdened with the cost of finding, developing, and producing the minerals.

Operating Agreements

A clear and comprehensive operating agreement is a necessary ingredient to any oil and gas deal. In most cases, several entities share the costs and revenues associated with an oil and gas lease. In these cases, an operator must be designated to carry out the terms and obligations of the lease for the parties. An operator, as discussed earlier, carries on the day-to-day management activities for the lease.

The operator's specific duties and limitations should be clearly specified. Model form operating agreements such as the A.A.P.L (American Association of Petroleum Landmen) Form 610 are available. The operating agreement should include discussion on cost and revenue sharing arrangements, operator responsibilities, expenditure limits, and nonconsent clause. The operating agreement will usually refer to the COPAS or accounting procedures exhibit. A COPAS is a set of standards developed by the Council of Petroleum Accountants Societies of North America.

INTERNATIONAL OIL BUSINESS

In the United States oil, gas, and other minerals are owned by the person, company or trust that owns the land with unrestricted rights of disposition or *fee simple*. However, this is not the case in most countries of the world. In other countries the person, company, or trust that has title to the land may only own the surface rights. What is below the surface, with the possible exception of water wells, belongs to the government either in whole or in part. Since a landowner in the U.S. owns the land and all things beneath it, they may lease or sell the mineral rights to another person or company for exploitation immediately or in the future. The landowner continues to own the surface rights, but if the lessee or owner of the mineral rights decides to drill a well or dig for minerals, they have the right to enter the property with the equipment necessary to build access roads and prepare sites for his operations, paying the landowner only for surface damages.

In many countries, U.S. oil companies often enter into agreements wherein they become partners with foreign national oil companies, or they act as service partners working with foreign governments or foreign corporations on a contractual basis. Because of the widespread differences in laws, cultures, language, and economic parameters found worldwide, it has been impossible to standardize international agreements. However, four basic categories of contractual agreements can be found internationally. The types of agreements are concession agreements, joint venture agreements, production sharing agreements, and service contracts. These contracts are discussed in Chapter 7.

Since the tax status of seller and buyer may be very different, this alone may be, and often is, the source of a substantial positive value gap. The tax treatment of farming varies considerably from country to country. In some countries, disposal of acreage may expose the seller to capital gains tax. In other countries, there is no such thing as a gains tax, and so on.

Government Approval

Since farming merely juggles the equity interests of private companies within an agreed concession, one might think that governments would have little or no involvement in the farming process. This is indeed the attitude of some governments whereas other governments may feel strongly about farming.

Documentation

The paperwork associated with farming can be fairly daunting, and it may explain why affiliates' legal groups always seem to be overworked. Requirements may vary from country to country, but typically five main documents are required.

- *The Farmout Agreement*. This has to be signed by seller and buyer, and it records the terms of the agreed deal. Upon satisfactory completion of all documentation, we will give you X% of concession ABC, and you will do the following work.
- *Request for Consent*. This is a letter to the government describing what is happening and asking for its consent. Depending on the JOA, a similar request may have to be sent to each partner.
- *Interest Assignment*. Although the concession assignment has put the buyer into the concession as far as the government is concerned, it has not established that the buyer gets any benefits under the JOA, i.e., is the buyer entitled to a share of the production. The interest assignment, signed by seller and buyer, does this.
- *Joint Operating Agreement Novation*. The JOA novation has to be signed by the buyer, the seller, and all partners. In it the buyer confirms to abide by the rules of the joint operating agreement and fulfil all its duties.

Competitive Invitation

In the international environment, acreage is usually awarded by governments on a discretionary basis that can be by individual initiative or by competitive invitation. Governments like to receive competing offers for acreage. To increase the likelihood of this, many governments solicit industry's views as to the acreage of interest and then invite applications for some or all of the proposed acreage.

CAPITAL MARKETS

The capital structure of a firm encompasses the right-hand side of the balance sheet, which describes how the firm's assets are financed. The permanent financing of the firm is represented primarily by:

- *Long-term Debt.* Most firms renew (roll over) their long-term obligations. Often, long-term debt is effectively permanent.
- *Preferred Stock.* Stocks that are sold to a unique group of shareholders who receive a monthly rate of return as stipulated in their stock certificates.
- *Common Shareholders' Equity.* Consists of three items: (1) capital received on the sale of the common stock, (2) additional paid-in capital over the par value of the stocks, and (3) retained earnings which is a result of the accumulation of the earnings and losses of the firm less dividends paid to the stockholder.

Long-Term Debt

The three most important classes of fixed income securities are term loans, bonds, and preferred stocks. These securities come in many types: secured and unsecured, convertible and nonconvertible, zero coupon and normal coupon and so on. The variety of types stems from the fact that different groups of investors favor different types of securities, and their tastes change over time.

Term Loans

A term loan is an agreement under which a borrower agrees to make interest payments and to repay principal, on specific dates, to a lender. The financial institution that lends the funds is usually a bank, insurance company, or pension fund. The maturity of a term loan is generally from three to fifteen years, but it may be as short as two or as long as thirty years. Most term loans are repaid gradually over the life of the loan agreement. This process is known as *amortization.*

Amortization calls for a series of equal installments over the life of the loan and requires the borrower to repay the loan on a regular basis and protects against default due to inadequate provision for repayment during the life of the loan.

Bonds

A bond is a long-term contract under which a borrower agrees to make a series of payments of interest and principal to the holder of the bond. Bonds differ from term loans in that they are generally offered to the public rather than to a single lender or a small syndicate of lenders. A bond's *indenture* sets forth the terms and conditions to which the bond is subject. The SEC approves indentures and makes sure that all indenture provisions are met before a company is allowed to sell new securities to the public. Some characteristics of bonds are:

- When real estate or other property is pledged in support of the bond issue, it is referred to as a *mortgage bond.*
- A *debenture* is an unsecured bond, and holders are general creditors of the corporation whose claims are protected by unpledged property.
- *Zero coupon bonds* are offered at a substantial discount below their par values. The advantages to the issuer are: (1) no cash outlays are required until maturity, (2) they often have a lower required rate of return than coupon bonds; and 3) the issuer receives an annual tax deduction. The advantages to the investors are (1) there is little danger of a *call*, and (2) zero coupon bonds guarantee a "true" yield to maturity since there is no reinvestment rate risk.
- *Convertible bonds* are securities that are convertible into a fixed number of shares of common stock at the option of the bondholder.
- Bonds issued with *warrants* provide the investor with an option to buy the common stock at a stated price.
- *Income bonds* pay interest only when permitted by the earnings of the firm.
- *Indexed bonds* have their coupon rates tied to an inflation index, such as the consumer price index.
- Bonds generally have maturities from 5-40 years when issued, although most have historically been in the 20- to 30-year range.
- A *call provision* gives the issuing corporation the right to call the bond issue for redemption at a predetermined price before its regular maturity.
- Bond issues are normally assigned *quality ratings* by both Moody's Investors Service and Standard & Poor's Corporation. The ratings reflect the probability that a bond issue will go into default. AAA and AA are the highest ratings signifying high quality and little chance of default. A- and BBB-rated bonds are of investment grade. Bonds with these ratings are the lowest-rated securities that many institutional investors are permitted to hold. Debt

rated BB and below is speculative; such bonds are *junk bonds*. CCC to D are very poor debt ratings. The likelihood of default is significant, or the debt is already in default.

Preferred Stock

Preferred stock represents an equity investment in a business firm, but it has many of the characteristics of a bond issue. Preferred stock usually has a par value, normally $25 or $100; however many recent issues have no par value. The preferred dividend is normally fixed in amount and stated either as a percentage of par value or in dollars.

Characteristics of preferred stock are:

- Most preferred stocks are cumulative. This means that if any preferred dividends have not been paid, this arrearage must be paid before any dividends can be paid on the common stock.
- Preferred stock generally has no maturity date, but it is often convertible into common stock at the option of the preferred stockholder (convertible preferred).

Advantages and disadvantages of bonds and preferred stocks include the following:

- The fixed income limitation of bond and preferred stock is a disadvantage to investors, but it is an advantage to corporations. Income from bonds and preferred stock are based on a given percentage rate of interest or dividend.
- Financing with bonds and preferred stock will not reduce current stockholders' control of the firm.
- Bond interest is deductible for income tax purpose while preferred and common dividends are not. This is an important incentive to raise capital via bonds rather than through a stock issue.
- From the issuer's viewpoint, bonds and term loans, with their required payments and restrictive provisions, are more risky than preferred or common stock. However, the reverse is true for the investor.
- The indenture provisions of a bond issue are normally much more restrictive in terms of limiting future financing than are provisions relating to issues of preferred or common stock.

Common Stock

Common stock represents the ownership of an incorporated business. It corresponds to the proprietor's capital or the partners' capital for an unincorporated business. In a corporation, common stockholders have control of the firm through their election of the firm's directors, who in turn select officers to manage the business. In a large, publicly owned firm, neither the managers nor any individual shareholders normally have the 51% necessary for absolute control of the company. Thus, stockholders must vote for directors, and the voting process is regulated by both state and federal laws.

LONG-TERM FINANCING DECISION OF AN INTERNATIONAL COMPANY

The long-term financing decision of the MNOC involves some aspects similar to short-term financing. Recall that the "effective" cost of short-term financing considered both the quoted interest rate and the percentage change in the exchange rate of the currency borrowed over the loan life. Just as currencies exhibit different interest rates on short-term bank loans, bond yields can vary as well among currencies.

Because Swiss and German bonds sometimes have lower yields or interest rates, it should not be surprising that U.S. corporations often consider issuing bonds in those countries denominated in Swiss francs or German marks.

To make the long-term financing decision, the firm must (1) determine the amount of funds needed, (2) forecast the price at which it can sell the bond, and (3) forecast periodic exchange rate values for the currency it plans to use for denominating the bond. This information can be used to determine the bond's financing costs, which can be compared with the financing costs the firm would incur using its home currency. Finally, the uncertainty of the actual financing costs to be incurred from foreign financing must be accounted for as well.

Floating-Rate Eurobonds—An Additional Risk to Consider

Eurobonds are often issued with a floating, rather than fixed, coupon rate. This means the coupon rate will fluctuate over time in accordance with other market interest rates. For example, the coupon rate may somehow be tied to the London Interbank Offer Rate (LIBOR) which is a rate at which Eurobanks lend funds to each other. As LIBOR increases, so would the coupon rate of a floating-rate bond. A floating coupon rate can be an advantage to the bond issuer during periods of decreasing interest rates, when otherwise the firm would be locked in at a higher coupon rate over the life of the bond. It can also be a disadvantage during periods of rising interest rates.

Long-Term Financing in Multiple Currencies

In some cases, the appropriate selection may be not a single currency or bond, but a portfolio of currencies for a borrower. Because the lifetime of bonds is too long to single out any particular currency as being safe, a portfolio of diversified currencies could reduce the risk incurred by the bond issuer. For example, a U.S. firm may denominate bonds in several foreign currencies rather than a single foreign currency so that substantial appreciation of any one particular currency will not drastically increase the dollars necessary to cover the financing payments.

COST OF CAPITAL

Cost of capital is the realm where corporate management establishes investment guidelines based on how much it costs the company to finance its activities. The cost of capital depends on the cost of debt, the cost of equity, and the corporate *capitalization structure*. The capitalization structure of a company is essentially the corporate balance of equity (common stock) and debt financing. When financial analysts talk about *financial leverage* they are referring to the amount of debt financing a company uses. Theoretically there should be some ideal capital structure, say perhaps 40% debt, for a particular company or even for a given industry.

Part of the determination of the financial structure deals with the cost of debt financing and the cost of equity financing. A typical oil company may be paying 10% interest on its bonds, but paying only a 5% dividend on common stock, i.e., the dividend yield = 5%. The debt sounds more expensive at first.

Cost of Debt

The *cost of corporate* debt is usually from 1.5 to 2.5 percentage points above long-term government bond rates. Interest payments are deductible, so if a company is paying 34% tax for example, the actual cost of debt financing (after tax) is 66% of the 10% interest rate or 6.6%. This still sounds higher than the dividend payment.

Cost of Preferred Stock

There is no tax benefit for preferred dividends from the perspective of the issuing company. Preferred dividends are not tax deductible like interest expense. The cost of preferred stock capital is the dividend per share divided by the price per share less the cost of issuing the stock. The costs of issuing or *floating* preferred stock can range from 2 to 4%. For example, the dividend to price ratio for most preferred stocks is around 9%. If the issuing or underwriting costs are 4%, the cost is calculated at 9.37%. The cost of preferred stock capital is shown in Figure 9–8.

$$\text{Cost of Preferred Stock} = \frac{\text{Dividend}}{\text{Stock price} - \text{Cost of issuing}}$$

$$\text{Cost of Preferred Stock} = \frac{9\%}{100\% - 4\%}$$

$$= 9.37\%$$

Fig. 9–8 Cost of Preferred Stock Capital

Cost of Equity

Equity capital is usually more expensive than debt. In some *over-leveraged* companies, debt is such a burden that the cost of debt approaches the cost of equity. Cost of equity can be viewed a couple of different ways.

Some analysts take the dividend yield of a stock and add to that the expected growth rate of the dividend stream. A company paying a 5% dividend that is expected to grow at 5% would have a cost of equity of 10%. Some stocks do not pay a dividend and a measure of the cost of equity based on earnings is usually preferred. Earnings per share is divided by the stock price and added to the expected growth rate of the stock as shown in Figure 9–9.

$$\text{Cost of Equity} = \frac{\text{Earnings per Share}}{\text{Stock Price}} + \text{Growth rate}$$

Fig. 9–9 Cost of Equity

A stock trading at 12 times earnings that is expected to grow at a rate of 5% per year would have a cost of equity equal to the earnings yield of 8.3% plus the 5% growth rate. This gives a cost of equity of 13.3%. This is over twice as costly as the 6.6% debt financing under these assumptions.

Capital Asset Pricing Model

The Capital Asset Pricing Model (CAPM) is a more sophisticated and accepted method for estimating the cost of equity. It is also used to determine the discount rate that should be used to evaluate a stock. It is based on the assumption that investors must aim for higher returns when dealing with the higher risks in the stock market. The CAPM calculates the cost of equity based on a risk-free return such as an U.S. government bond, plus an adjusted risk premium for the particular stock. The adjusted risk premium is based on the market rate of interest and the beta of the stock.

Market Rate of Interest

Two basic elements make up the market rate of interest, or the market rate of return. The first is the relatively risk free rate of interest of an U.S. government bond--about 8.5% which is composed of a real interest rate component and an inflation component. The real rate of interest is calculated by subtracting the inflation rate from the quoted nominal interest rate. The second is the risk premium investors require to justify being involved with equity securities. Historically, market premiums have ranged from 4% to 7%. The relationships are shown in Figure 9–10.

Risk Free Rate	=	Real interest rate	3.5%	
	+	Inflation component	3.5%	
	=	Government bond	7.0%	Nominal rate
	+	Risk premium	5.5%	
	=	Market rate of interest	12.5%	

Fig. 9–10 Components of the Market Rate of Interest

Beta

The Beta of a stock measures its trading price volatility relative to either a stock market index or an industry related index of stocks. If a stock's price tends to follow its industry group up or down in synchronization, the stock will have a Beta of 1. A stock that rises more than other stocks in a bull market and falls faster in a bear market will have a Beta greater than one. A high Beta stock will exhibit a more volatile performance during market fluctuations. If every time the market went up 10% the stock of Company X would go up 12%, the Beta for the company relative to the market would be 120% (or 1.2).

The Beta, market rate of interest, and the risk-free rate of interest are used to calculate the cost of equity capital for a company. An investor would use the same information to calculate his *required rate of return* for investing in a stock with the same parameters. The Value Line of betas for several oil and gas companies are shown in Table 9–2.

	4/1988	4/1990	1/1993	3/1995
Unocal	1.05	1.05	1.00	.95
Mobil	.90	.85	.75	.65
Oxy	.95	.90	.95	.80
Phillips	.90	1.00	.90	.80
Amoco	.80	.80	.75	.65
ARCO	.85	.85	.75	.70
Chevron	.95	1.00	.85	.70
Total	.90	.95	NMF	.60
Texaco	.75	.75	.65	.65
RD Shell	.80	.75	.70	.70
Average	.89	.89	.73	.72

Table 9–2 Value Line—Betas

CAPITAL ASSET PRICING MODEL

The capital asset pricing model for calculation of the required rate of return on common equity is shown in Figure 9–11.

RRR = Rf + Bi(Rm – Rf)

where

RRR = Required rate of return from investor point of view, or cost of equity capital
Rf = Risk free rate of return (U.S. government bond)
Rm = Market rate of return
Bi = Beta of the investment
Bi(Rm – Rf) = Risk Premium for a particular stock with a Beta equal to Bi.

RRR = 7.0% + 1.25 (14.0% – 7.0%)

RRR = 7.0% + 8.75%

RRR = 15.75%

Fig. 9–11 Capital Asset Pricing Model

Weighted Average Cost of Capital

Many analysts prefer to determine discount rates, reinvestment rates and company cost of capital by using the *weighted average cost of capital.* The cost of each component of corporate financing is weighted according to its percentage of the capital structure.

The example here is Company X with a beta of 1.25. The company has a capital structure that consists of 30% debt and 10% preferred stock. The after-tax cost of debt is 6.6%, and the 10% of capital provided by preferred stock has a cost of 9.0% and 5% deferred taxes at zero (ø) cost. The CAPM calculates the cost of equity at 15.75%. The overall cost of capital is summarized below using the weighted average cost of capital (WACC) approach. Each form of corporate financing is *weighted* according to its market value percentage relative to the total *market capitalization* of the company. An example calculation is illustrated in Table 9–3.

Source of Capital	Cost %	Weight %	Weighted %
Debt Financing	6.6%	30%	1.98%
Preferred Stock	9.0%	5%	.45%
Deferred Taxes	0%	5%	0%
Equity Financing	15.75%	60%	9.45%
Weighted Composite Average		100%	11.88%

Table 9–3 Weighted Composite Cost of Capital

The weighted average of 11.88% represents the cost of capital for Company X. The company would theoretically not invest in any venture that yielded an after-tax internal rate of return (IRR) of less than say 12%. For growth rate of return (GRR) calculations, the company would probably use a reinvestment rate of 12%. There are other considerations of course, but this is the benchmark for determining the boundary conditions for corporate financing and investment policy.

This is the common example used in presenting the concept of cost of capital, but determining the cost of capital has elements of scientific procedure and art. Estimating the market rate of interest, for example, can be quite subjective. Furthermore, the position held by *deferred taxes* in the corporate capital structure can be fairly abstract. It is usually considered to be the equivalent of an interest-free loan from the government. It normally does not amount to a substantial portion of the total capitalization of a firm, but this is the most common treatment if it is factored in at all.

Growth Rate

In all the equations for present value, the key factors are interest or discount rates, and growth. There are many ways of estimating earnings or cash flow growth. One of the most common is to calculate the growth rate of earnings or cash flow over a period of time. If earnings appear to increase consistently at a rate of 10% per year, then an analyst might be able to start with that and make some projections. Anyone in the oil industry knows that this is less realistic for oil companies than for other industries. The oil industry lives with volatile prices. One calculation of the rate of growth for a company is based on the formula Figure 9–12.

$$\text{Growth Rate} = \frac{\text{Net Income} - \text{Dividends}}{\text{Shareholder Equity}}$$

Fig. 9–12 Calculation of Equity Growth Rate

Companies that pay out a larger share of net income in dividends would theoretically have a relatively smaller growth rate. One reason why growth companies characteristically pay no dividends can be seen in this formula. The lower the dividend rate, the higher the growth rate.

From a financial point of view this would be considered a measure of accounting growth rather than economic growth. The formula must rely on accounting measures of corporate value (shareholder equity) and the increase in corporate wealth (net income after dividends). This approach is considerably less appropriate in the oil industry than many other industries. The differences between economic value and accounting value are greater in the oil industry than most other industries.

THE IMPORTANCE OF THE CAPITAL STRUCTURE AND FINANCING

A great deal of controversy has developed over whether the capital structure of a firm, as determined by its financing decision, affects its cost of capital. Traditionalists argue that the firm can lower its cost of capital and increase market value per share by the judicious use of leverage. However, as the company levers itself and becomes increasingly risky financially, lenders begin to charge higher interest rates on loans.

Moreover, investors penalize the price/earnings ratio increasingly with higher leverage, all other things being the same. Beyond a certain point, the cost of capital begins to rise.

According to the traditional position, that point denotes the optimal capital structure. However, there are those who argue that in the absence of corporate income taxes, and deductibility of interest, the cost of capital is independent of the capital structure of the firm. They contend that the cost of capital and the total market value of the firm are the same for all degrees of leverage. With the introduction of corporate income taxes, debt has a tax advantage and serves to lower the cost of capital.

In deciding upon an appropriate capital structure, the financial manager should consider a number of factors. Considerable insight can be obtained from an analysis of the cash-flow ability of the firm to service fixed charges associated with senior securities and leasing. By evaluating the probability of cash insolvency, one is able to determine the debt capacity of the firm.

The greater the dollar amount of senior securities the firm issues and the shorter their maturity, the greater the fixed charges of the firm. These charges include principal and interest payments on debt, lease payments, production payments, and preferred stock dividends. The greater and more stable the expected future cash flows of the firm, the greater the debt capacity of the company.

From an internal standpoint, financial risk associated with leverage should be analyzed on the basis of the firm's ability to service fixed charges. This analysis should include the preparation of cash budgets to determine whether the expected cash flows are sufficient to cover the fixed obligations.

Another method is to analyze the relationship between EBIT and earnings per share (EPS) for alternative methods of financing. This analysis is then expanded to consider likely fluctuations in EBIT and light is shed on the question of financial risk. To illustrate, suppose that a firm wished to compare the impact on earnings per share financing a $10 million expansion program either with common stock at $50 a share or with 8% bonds. The tax rate is 50%, and the firm currently has an all-equity capital structure consisting of 800,000 shares of common stock. At $50 a share, the firm will need to sell 200,000 additional shares in order to raise $10 million. At a hypothetical EBIT level of $8 million, earnings per share for the two alternatives the calculation of EPS is shown in Table 9–4.

	Common-Stock Financing	Debt Financing
EBIT	$8,000,000	$8,000,000
Less Interest	0	800,000
EBT	$8,000,000	$7,200,000
Less Taxes	4,000,000	3,600,000
Earnings after Taxes	$4,000,000	$3,600,000
Shares Outstanding	1,000,000	800,000
EPS	$4.00	$4.50

Table 9–4 Calculation of Earnings Per Share (EPS)

Also, the financial manager can learn much from a comparison of capital structure ratios for similar companies. Using a comparison, the companies should be in the same industry, i.e., a comparison for an oil company should be with other oil companies that have the same type of operations. An integrated company should compare with another integrated company; a refinery operation should be compared with another refinery operation, etc. Allowances for different accounting practices should also be considered.

Preferably, a full-cost exploration and production (E&P) company should be with other full-cost E&P companies or make adjustments to compensate for the accounting differences. If the firm is contemplating a capital structure significantly out of line with that of similar companies, it is conspicuous to the marketplace. The optimal capital structure for all companies in the industry might call for a higher proportion of debt to equity than the industry average. As a result, the firm may well be able to justify more debt than the industry average.

Other comparisons can be made using regression studies and simulations. The importance of regression analysis and simulation techniques in providing valuation information to the financial manager should not be undervalued, but it is important to point out the practical limitations of the models. Probability information usually is based upon empirical testing of valuation models, the results of which have been far from precise or consistent. An equally thorny problem is the specification of the relationship between variables.

The firm may also profit by having discussions with investment analysts, investment bankers, and lenders to determine their views on the appropriate amount of leverage. These analysts examine many companies and are in the business of recommending stocks. Therefore, they have an influence upon the market, and their judgments with respect to how the market evaluates leverage may be very worthwhile.

Collectively, the methods of analysis should provide sufficient information on which to base a capital structure decision. Once an appropriate capital structure has been determined, the firm should finance investment proposals in roughly those proportions.

Financial Reporting and Investor/Stockholder Relations

10

In recent years the petroleum industry has been the subject of seemingly endless comment, debate and criticism. The entire world has become aware of the inordinately significant impact of the activities and actions of oil and gas companies. An issue of much debate has been the financial accounting and disclosure practices of oil and gas producing companies. Most users of financial information feel the reports developed from general accepted accounting principles (GAAP) did not reflect the true market value of the petroleum company's assets. Investors could not determine if their investment in the company would stand the test of the marketplace over time.

Oil and gas companies are keenly aware of the importance of communicating to the public. Communications to the public regarding economic and social issues impact the public's view of the corporation. The public's reactions, whether positive or negative, affect the marketplace and consequently, the company's earnings and market value. The focus of this chapter is on how the oil and gas companies have responded regarding financial reporting and maintaining positive relations with the investing public.

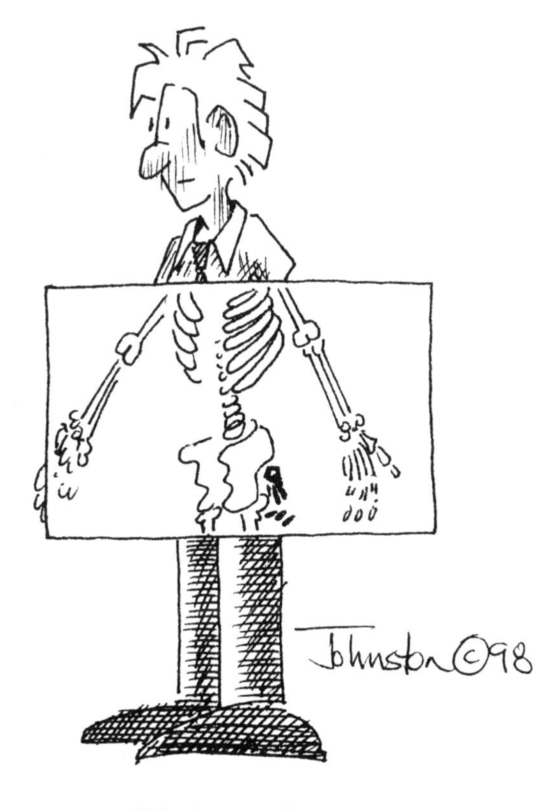

"Disclosures"

DEVELOPMENT OF SFAS 69

The most valuable assets of an oil or gas company, oil and gas reserves, are neither visible nor recorded on the firm's accounting records. Before 1982, both the Security Exchange Commission (SEC) and the Financial Accounting Standards Board (FASB) considered several procedures to adequately report the actual value of oil and gas assets. In the absence of uniform procedures and inability of the SEC and FASB to agree upon a standard set of guidelines, oil and gas producing companies employed a variety of reporting techniques. All guidelines were designed to comply with generally accepted accounting principles (GAAP) and overall reporting norms.

In 1975, acting on a wave of public sentiment and reacting to intermittent shortages of petroleum products, the U.S. Congress passed the Energy Policy and Conservation Act of 1975 (EPCA). This legislation paved the way for the current proliferation of information demands placed upon oil and gas producing companies by various public and private accounting standard-setting authorities.

Financial accounting, disclosure requirements and reporting standards have grown from an informal set of rules to a voluminous collection of multiple disclosure requirements. These reporting standards are previously unheard of in any single U.S. industry segment. In addition, the FASB and the SEC have expanded disclosure requirements to unchartered regions on the very fringes of the current accounting conceptual framework.

Oil and gas companies and the accounting profession have primarily been unenthusiastic over the inclusion of reserve quantities, and especially over the inclusion of reserve valuations in published financial statements. The difficulty of estimating reserve quantities is often cited in arguments against such disclosures. The accounting profession's skepticism over the desirability of reserve disclosures is influenced by their conservative viewpoint.

Reserve Recognition Accounting (RRA)

Many methods have been considered to find a way to adequately represent the actual value of oil and gas assets. In 1978, in response to a request from the SEC, the FASB announced a program of financial reporting, FASB No. 19, termed Reserves Recognition Accounting (RRA). The objective was to provide for the *value* of a company's reserves to be *recognized* as an asset. Also, additions to proved reserves would be recognized as an asset and the additions included in earnings.

The SEC originally intended RRA to replace Full Cost (FC) and Successful Efforts (SE) accounting methods. But RRA was only required as supplemental information during a trial period from January 1979 to November 1982. The FASB issued statement No. 25 in February 1979 suspending all but the disclosure requirements of FASB No. 19. It was decided that RRA could not replace FC and SE accounting due primarily to the inaccuracies of reserve reporting. However, the SEC made it clear that reserve valuation is an important element in the disclosure package. They said if the FASB did not include value-based reserve disclosures in its requirements, they would mandate such a requirement.

The FASB finally chose a supplemental reporting system and on September 28, 1982 issued SFAS No. 69. The SEC relinquished their proposed Reserve Recognition Accounting (RRA) guidelines and agreed with the FASB's supplemental reporting guidelines for fiscal years beginning after December 15, 1982.

The main result is "the standardized measure of oil and gas" or SMOG. It is nearly identical to what is sometimes called SEC value of reserves or RRA value. Companies required to present this information are publicly traded companies with significant oil and gas producing activities that meet one or more of the following criteria specified in SFAS No. 69:

1. Revenues from oil and gas producing activities (including both sales to unaffiliated customers and sales or transfers to the enterprise's other operations) are 10% or more of the combined revenues of all the enterprise's industry segments.
2. Results of operations for oil and gas producing activities, excluding the effect of income taxes, are 10% (or more) of the greater of:
   a. The combined operating profit of all industry segments that recognized a profit.
   b. The combined operating loss of all industry segments that recognized a loss, or
   c. The identifiable assets relating to oil and gas producing activities are 10% or more of the combined identifiable assets of all industry segments.

REQUIRED SUPPLEMENTAL DISCLOSURES

SFAS No. 69 requires that publicly traded companies with significant oil and gas producing activities disclose supplementary information in the annual financial statements related to the following items:

Historical-Based Reports

1. Proved reserve quantity information
2. Capitalized costs relating to oil and gas producing activities
3. Costs incurred for property acquisition, exploration, and development activities
4. Results of operations for oil and gas producing activities

Future Value Reports

5. A standardized measure of discounted future net cash flows relating to proved oil and gas reserve quantities
6. Changes in standardized measure of discounted future net cash flows relating to proved oil and gas reserve quantities

Successful Efforts Method GAAP Procedures	Reserve Recognition Supplemental Disclosures	Full Cost Method GAAP Procedures
•Capitalizes only cost of successful exploration, drilling and development projects	•Proved Reserve quantity information •Capitalized costs relating to oil and gas producing activities •Costs incurred for property acquisition, exploration, and development activities •Results of operations for oil and gas producing activities •Standardized measure of discounted future net cash flows relating to proved oil and gas reserve quantities •Changes in discounted future net cash flows	•Capitalizes cost of successful exploration, drilling and development projects plus •Costs incurred for dry holes

Figure 10–1 Comparison of Valuation Methods

All six disclosures must be presented in the aggregate. Disclosures 1, 2, 3, and 5 must also be presented for each geographical area. Disclosures 1–4, the historical-based disclosures, are relatively uncomplicated and require little explanation in addition to the examples.

Disclosures 1 and 3 are the same for both FC and SE companies while Disclosures 2 and 4 differ for FC and SE companies. Disclosures 5 and 6, the future-based disclosures, are much more complicated and require significantly more computations than the historical-based disclosures.

Proved Reserve Quantity (Disclosure 1) Contents Include:

1. Quantification of net proved oil and gas reserves attributable to the enterprise's interest
2. Reserves are categorized as *developed* or *undeveloped*, and
3. Changes in the net quantities of the enterprise's proved oil and gas reserves during the year

The reserves reported should include both working interests and nonworking interests owned. The reserve quantity disclosure reports only on reserve quantities. No costs, either capitalized or expensed, are reported. Consequently, the disclosure would be the same regardless of successful efforts or full cost methods of accounting. (See Table 10–1)

Oil and Gas Reserves - 1995

(Gas in billions of cubic feet Crude Oil and NGL in million of barrels)	Worldwide		
	Crude Oil	**NGL**	**Gas**
Proved Reserves			
December 31, 1994	1,696	509	18,521
Revisions of previous estimates	25	12	(86)
Improved Petroleum applications	105	4	291
Extensions, discoveries and other activities	146	13	,803
Purchases of reserves in place(1)	69	4	341
Sales of reserves in place	(35)	(4)	(180)
Production	(192)	(30) [2]	(1,537)
December 31, 1995	**1,814**	**508**	**19,153**
Proved Developed reserves			
December 31,1994	1,455	459	15,538
December 31,1995	1,458	460	15,441

[1] In 1995, purchases of reserves in place include 56 million barrels associated with interest in Azerbaijan.

[2] Excludes nonleasehold NGL production attributable to processing plant ownership of approximately 15 million barrels for 1995.

Table 10–1 Supplemental Disclosure No. 1 for Oil and Gas Companies. The actual format includes detail for years 1993 and 1994. Also geographical breakdown for United States, Canada, Europe and Other was presented for each period.

PROVED OIL AND GAS RESERVES. Proved oil and gas reserves are the estimated quantities of crude oil, natural gas, and natural gas liquids that geological and engineering data demonstrate with reasonable certainty to be recoverable in future years from known reservoirs. Estimates are made under existing economic and operating conditions, i.e., prices and costs as of the date the estimate is made. Prices include consideration of changes in existing prices provided only by contractual arrangements, but not on escalations based upon future conditions.

MINERAL INTERESTS IN PROPERTIES. Including (1) fee ownership or a lease, concession or other interest representing the right to extract oil or gas subject to such terms as may be imposed by the conveyance of that interest; (2) royalty interests, production payments payable in oil or gas, and other nonoperating interests in properties operated by others; and (3) those agreements with foreign governments or authorities under which a reporting of ownership in reserves is made.

Changes in the net quantities of the enterprise's proved oil and gas reserves during the year include:

- Revisions of previous estimates. Revisions represent changes in previous estimates of proved reserves, either upward or downward, resulting from new information (except for an increase in proved acreage) normally obtained from development drilling and production history or resulting from a change in economic factor.
- Improved Recovery. Changes in reserve estimates resulting from application of improved recovery techniques shall be shown separately, if significant. If not significant, such changes are included in revisions of previous estimates.
- Extensions and discoveries. Additions to proved reserves that result from (1) extension of the proved acreage of previously discovered (old) reservoirs through additional drilling in periods subsequent to discovery, and (2) discovery of new fields with proved reserves or of new reservoirs of proved reserves in old fields.
- Production. Actual production during the period, and
- Sales of Minerals in place. Sales of ownership of proved properties to other producers.

Capitalized Costs Relating to Oil and Gas Producing Activities (Disclosure 2)

The report includes the aggregate amount of capitalized costs relating to an enterprise's oil and gas producing activities and the aggregate related accumulated depreciation, depletion, amortization, and valuation allowances. Costs are reported as incurred by geographic areas. If significant, capitalized costs of unproved properties are reported. Capitalized costs of support equipment and facilities may be disclosed separately or with capitalized costs of proved and unproved properties. Investments accounted for by the equity method are included and reported separately. (Table 10–2)

Costs Incurred and Capitalized

(millions of dollars)	United States	Canada	Europe	Other	World-wide
Year 1955					
Property Acquisition					
Proved	$ 176	$ 6	$ -	$ -	$ 182
Unproved	74	33	-	28	135
Exploration	262	124	179	409	974
Development	769	288	344	306	1,707
Total	**$1,281**	**$ 451**	**$ 523**	**$ 743**	**$2,998**
Year 1954					
Property Acquisition					
Proved	$ 52	$ 11	$ 9	$ 1	$ 73
Unproved	$ 50	$ 51	$ 3	$ 2	$ 106
Exploration	$ 245	$ 116	$ 185	$ 291	$ 837
Development	$ 614	$ 246	$ 193	$ 446	$1,499
Total	**$ 961**	**$ 424**	**$ 390**	**$ 740**	**$2,515**
Year 1953					
Property Acquisition					
Proved	$ 11	$ 11	$ 36	$ 36	$ 81
Unproved	$ 4	$ 23	$ 54	$ 20	$ 101
Exploration	$ 133	$ 64	$ 149	$ 229	$ 575
Development	$ 657	$ 234	$ 276	$ 427	$1,594
Total	**$ 805**	**$ 332**	**$ 515**	**$ 699**	**$2,351**

Table 10–2 Typical Format of Costs Incurred Disclosure

Disclosure of Costs Incurred (Disclosure 3)

Costs are reported for property acquisition, exploration and development whether capitalized or charged to expense at the time they are incurred. The amounts are disclosed separately for each of the geographic areas for which reserve quantities are disclosed. (Table 10–3)

Average Sales Prices and Production Costs
Per Unit of Oil and Gas Produced –1995

(millions of dollars)		1995	1994	1993
Product Revenues				
Crude oil and natural gas liquids (dollars per barrel)				
United States	crude oil	$ 16.02	$ 14.82	$ 15.96
	natural gas liquids	$ 10.00	$ 9.39	$ 10.79
Canada	crude oil	$ 15.15	$ 13.38	$ 13.94
	natural gas liquids	$ 9.71	$ 8.75	$ 9.44
Europe		$ 17.18	$ 15.49	$ 17.69
Other		$ 16.02	$ 14.23	$ 15.87
Natural gas (dollars per mcf)				
United States		$ 1.35	$ 1.66	$ 1.88
Canada		$ .89	$ 1.39	$ 1.31
Europe		$ 2.49	$ 2.23	$ .81
Production Costs				
(dollars per equivalent barrel)(1)				
United States		$ 3.54	$ 3.89	$ 4.42
Canada		$ 3.29	$ 3.89	$ 3.27
Europe		$ 5.59	$ 6.62	$ 6.43
Other		$ 3.93	$ 3.84	$ 4.01

(1)Production costs are shown on a dollar-per-barrel basis after converting natural gas into equivalent barrel units. Natural gas was converted on the basis of approximate relative energy content.

Table 10–3 Disclosure No. 3 for Oil and Gas Companies

Results of Operations (Disclosure 4)

This includes, by geographical area, revenues less production (lifting) costs, exploration expenses, depreciation, depletion, and amortization, and valuation provisions. Income taxes are computed using the statutory tax rate for the period. Revenues include sales to unaffiliated enterprises and sales or transfers to the enterprise's other operations. Income taxes are computed using the statutory tax rate for the period. (Table 10–4)

Results of Operations for Oil and Gas Producing Activities - 1995

(millions of dollars)	United States	Canada	Europe	Other	World-wide
Oil and Gas Production Revenues					
From consolidated subsidiaries	$2,223	$ 331	$ –	$ 908	$3,462
From unaffiliated entities	512	274	719	717	2,222
Other Revenues	165	100	102	92	449
Total Revenues	2,890	705	821	1,717	6,133
Production Costs					
Taxes other than income	179	13	25	112	329
Other production costs	744	240	233	369	1,586
Exploration Expenses	152	112	123	223	610
Depreciation, Depletion and Amortization Expense	973	350	197	337	1,857
Other related costs	321	73	85	117	596
Total Costs	2,369	788	663	1,158	4,978
Operating Profit	521	(83)	158	559	1,155
Income Tax Expense	15	(37)	70	314	362
Results of Operations	$506	$(46)	$88	$245	$793

Table 10–4 Supplemental Disclosure No. 4 for Oil and Gas Companies. Actual report also includes detail for years 1993 and 1994.

Standard Measure (SMOG) (Disclosure 5)

This disclosure is based on standardized discounted cash flow analysis of proved reserves. (Table 10–5) The SMOG requirements that provide the basis of standardization are as follows:

a. Prices received at fiscal year-end for products (oil, gas, coal, and sulfur) sold

b. Prices are held constant, no escalation

c. Costs are not escalated

d. Future income taxes

e. A 10% discount rate is used

f. Standardized measure of discounted future net cash flows. This amount is the future net cash flows less the computed discount.

Standardized Measure of Discounted Future Net Cash Flows Relating to Proved Oil and Gas Reserves – 1995

(millions of dollars)	United States	Canada	Europe	Other	World-wide
December 31,1995					
Future Cash Inflows	$33,326	$ 7,534	$ 9,671	$13,359	$62,890
Future Development and Production Costs	15,923	3,759	4,174	5,173	29,029
Future Income Taxes	4,438	1,515	1,841	3,401	11,195
Future Net Cash Flows	12,865	2,260	2,658	4,785	22,666
Ten % Annual Discount	7,385	653	948	1,844	10,830
Discounted Cash Flow	**$ 5,580**	**$ 1,607**	**$ 1,708**	**$ 2,941**	**$11,836**

Table 10–5 Supplemental Disclosure No. 5 for Oil and Gas Companies. The actual format includes data for years 1994 and 1993.

Changes in Standard Measure and Reasons for Changes (Disclosure 6)

The changes include:

- Accretion of discount
- Sale of oil and gas produced, net of production costs
- Development costs (previously estimated incurred during the period)
- Prices and production costs
- Revision of previous quantity estimates
- Changes in estimated future income taxes
- Sales and purchases of oil and gas in place
- Other

Other Disclosures. The accounting method used must be disclosed as well as the manner of disposing of capitalized costs as a note to the financial statements. In addition to the reserve and financial disclosure requirements, the SEC requires a number of other elements to be included:

a. Average sales price and production cost per unit.

b. Total net productive and dry wells drilled broken down by geographic area and exploration vs. development.

c. Total gross and net productive oil and gas wells.

d. Total gross and net developed acreage by major geographical area.

e. Total gross and net undeveloped acreage by major geographic area and minimum remaining terms on leases if material.

f. Number of wells in progress, gross and net, and water floods, pressure maintenance operations, etc., and

g. Information about future obligations to provide fixed quantities of oil and gas under existing contracts.

Statement of Changes in Standardized Measure of Discounted Future Net Cash Flows – 1995

(millions of dollars)	**1995**	**1994**	**1993**
Balance at January 1	$10,991	$10,044	$12,670
Changes resulting from:			
Sales and transfers of oil and gas produced, net of production costs	(3,769)	(3,765)	(3,965)
Net changes in prices, and development and production costs	407	1,059	(3,966)
Current year expenditures for development	1,707	1.499	1.594
Extensions, discoveries and improved recovery, less related costs	1,922	1,128	758
Purchases (sales) of reserves in place	128	(46)	(235)
Revisions of previous quantity estimates	56	303	488
Accretion of discount	1,441	1,331	,798
Net change in income taxes	(833)	253)	1,861
Other	(214)	(310)	(959)
Balance at December 31	$11,836	$10,991	$10,044

Table 10–6 Supplemental Disclosure No. 6 for Oil and Gas Companies

USING THE INFORMATION FROM SFAS 69

There are a number of uses of the standardized measure SMOG information. It certainly adds to the body of information about a particular company regardless of its shortcomings.

1. SMOG provides a basis for estimating fair market value (FMV) of a company's reserves. It also provides a basis for comparison with other companies.
2. The SMOG values provide the basis for a variety of value added (EVA) analysis. The SMOG information allows a couple of perspectives on whether or not a company is in fact enhancing corporate wealth. The measure of corporate wealth growth is based upon the 10% discount rate required by the SEC. Of course, there are other boundary conditions such as the use of year-end prices with no price escalation. However, this measure provides consistency and some objectivity.
3. The SMOG information also provides an important dimension to the age-old statistics regarding reserves additions.

Average Sales Prices and Production Costs
Per Unit of Oil and Gas Produced – 1995

(millions of dollars)		1995	1994	1993
Product revenues				
Crude oil and natural gas liquids (dollars per barrel)				
United States	crude oil	$ 16.02	$ 14.82	$ 15.96
	natural gas liquids	10.00	9.39	10.79
Canada	crude oil	15.15	13.38	13.94
	natural gas liquids	9.71	8.75	9.44
Europe		17.18	15.49	17.69
Other		16.02	14.23	15.87
Natural gas (dollars per mcf)				
United States		1.35	1.66	1.88
Canada		.89	1.39	1.31
Europe		2.49	2.23	.81
Production Costs				
(dollars per equivalent barrel)[1]				
United States		3.54	3.89	4.42
Canada		3.29	3.89	3.27
Europe		5.59	6.62	6.43
Other		3.93	3.84	4.01

[1] Production costs are shown on a dollar-per-barrel basis after converting natural gas into equivalent barrel units.
Natural gas was converted on the basis of approximate relative energy content.

Table 10–7 Disclosure for Oil and Gas Companies

INVESTOR/STOCKHOLDER RELATIONS

When the company makes the decision to raise additional capital and determines that the best procedure is to raise capital in the capital market, the next step is to determine whether they want to raise the funds through private placements or through public placement. A firm's first issuance of securities to the public is an initial public offering (IPO). The process by which a closely held corporation issues new securities to the public is called "going public." When a firm goes public, it issues its securities as a new issue or initial public offering.

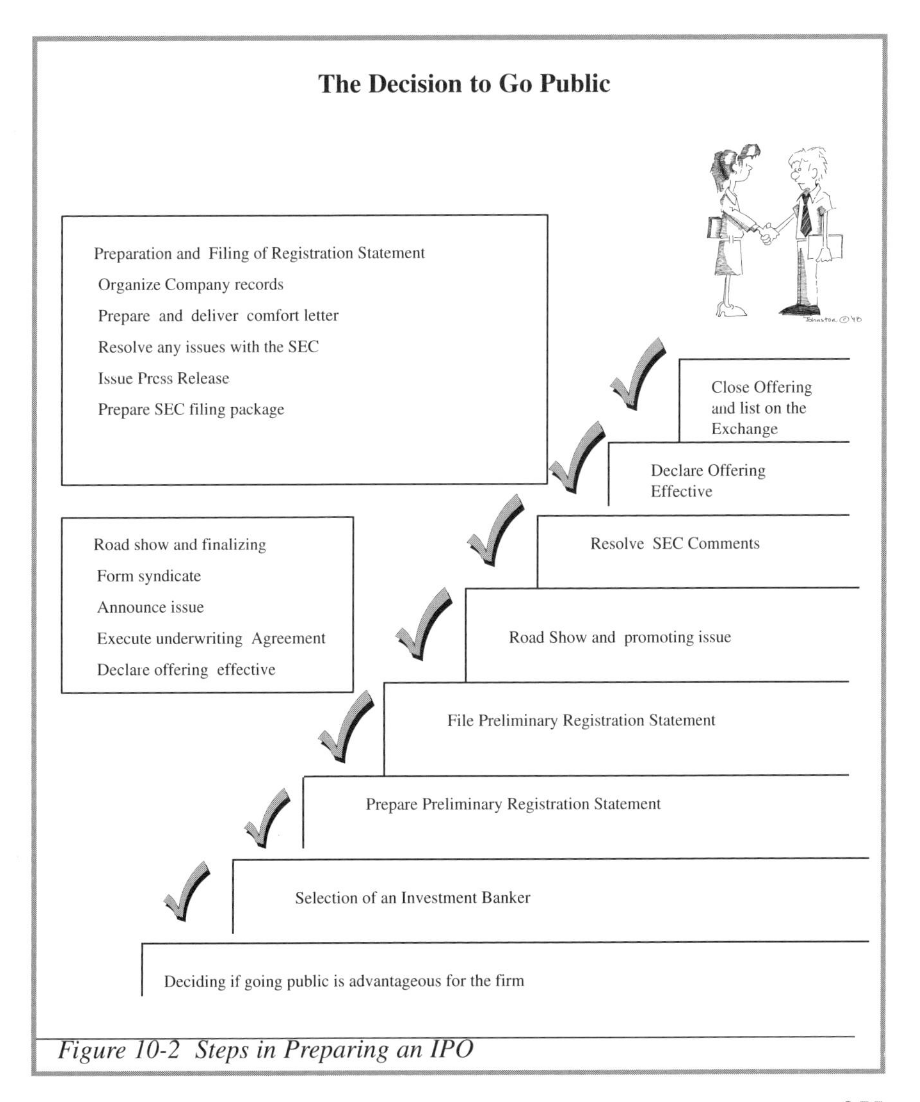

Figure 10-2 Steps in Preparing an IPO

Advantages of going public include:

- The ability to raise additional funds
- The establishment of the firm's value in the market, and
- An increase in the liquidity of the firm's stock

Disadvantages of going public include:

- Costs of the reporting requirements of the SEC and other agencies
- Access to the company's operating data by competing firms
- Access to net worth information of major shareholders
- Limitations on self-dealing by corporate insiders, such as officers and major shareholders
- Pressure from outside shareholders for earnings growth
- Loss of control by management as ownership is diversified
- Need for improved management control as operations expand, and
- Increased shareholder servicing costs

Investment Bankers

Investment bankers serve as intermediaries between businesses and the providers of capital. Moreover, they not only help to sell new securities but also assist in business combinations, act as brokers in secondary markets, and trade for their own accounts.

In their traditional role in the sale of new securities, investment bankers help determine the method of issuing the securities and the price to be charged, distribute the securities, provide expert advice, and perform a certification function. A single investment banker ordinarily does not underwrite an entire issue of securities unless the amount is relatively small. To share the risk of overpricing the issue or of a market decline during the offering period, the investment banker (the lead or managing underwriter) forms an underwriting syndicate with other firms.

The first step in selling the stock takes place when the investment bankers take the issue "on the road." This involves visits to the top management of the firm, the underwriter, many investment bankers, and institutional investors to convince them to buy a portion of the initial issue. This pre-selling helps to develop market for the stock and is essential to ensure that the entire issue can be sold.

Flotation Costs

The costs associated with a new security issue are termed flotation costs. These costs include compensation to the investment banker plus legal, accounting, printing, and other costs borne by the issuer. Flotation costs depend on the type of security issued and the size of the issue.

These costs, expressed as a percentage of gross sales, range from 1.1 to 14% for bonds, 1.6 to 2.6% for preferred stock, and 3.5 to 22% for common stocks. Flotation costs are relatively lower for large than small issues, tend to be greater for common stock than for preferred stock; and higher for stocks than for bonds.

For new issues, the investment banker will normally try to maintain a market in the shares for some time after the public offering. This is done in order to provide liquidity for the shares and to maintain a good relationship with both the issuer and the investors who purchase the shares.

SEC Registration Statement

Once the decision to make an initial public offering has been made, the questions are similar to those for seasoned issues: the amount to be raised, the type of securities to sell, and the method of sale. The company's next step is to prepare and file a SEC registration statement (S series) and prospectus. The Securities Act of 1933 prohibits the offer or sale through the use of the mails or any means of interstate commerce of any security unless a registration statement for that security is in effect or an exemption from registration is applicable.

The registration process has three distinct periods:

1. During the prefiling period, the issuer may engage in preliminary negotiations and agreements with underwriters. Offers to buy or sell securities are prohibited during this period.
2. The waiting period begins when the registration statement is filed. During this time, selling a security is still illegal; however, making an oral offer to buy or sell a security is not. Information may be published during the waiting period in tombstone ads.
3. The post-effective period begins once a registration statement becomes effective. A prospectus must be furnished to any interested investor. Its purpose is to supply sufficient facts to make an informed decision.

A *private placement* is exempt from registration with the SEC. It involves the sale of securities to a very few parties, such as institutional investors. A private placement of stocks or bonds is allowed, but bond placements are more common. The typical private placement is made by selling the securities directly to a financial institution. Direct business loans with maturities of 15 years or less are called *term loans*. If they have longer maturities, they are considered private placements that are likely to have more restrictive covenants than public issues and are usually easier to renegotiate in case of default. Private placements typically have higher interest rates.

Listing the Firm's Securities

For many firms, the decision to go public with the firm's stock coincides with the decision to have the stock listed on one of the organized exchanges. Some publicly held firms do not list their stock on any exchange, but rather issue and trade their stock directly with their shareholders or, as is most common, have the stock traded in the over-the-counter (OTC) market.

The primary advantage of having a firm's stock listed on one of the organized exchanges is that the marketability of the stock should increase, the stock can be more widely traded and more investors have easy access to it.

The disadvantages of listing securities include the ongoing reporting requirements and restrictions on the types of stocks that can be traded. Some firms decide to remain in the over-the-counter market rather than list their stock on an organized exchange to eliminate the costs of reporting.

Maintaining Relations with Stockholders and Investors and Public Image

Investor relations have been and still are an evolving activity. About 30 years ago, it often was a part of the public relations department, perhaps viewed as a somewhat specialized communication function. In the 1950s through the 1960s and into the early 1970s, with the rapid growth of employee retirement funds, many blue-chip companies used skilled communicators to educate portfolio managers and brokerage intermediaries about the investment merits of the company's stock.

The Council of Institutional Investors was founded in 1985 in response to controversial takeover activities threatening the financial interest of pension-fund beneficiaries. The Council counts among its most active member, the $78.5 billion California Public Employees Retirement System, the $48.5 billion State Teachers Retirement System as well as the $35.4 billion New York City Pension Funds. Because of the size of its investing power, the Council is quite effective in attaining their agendas.

Corporate Governance

And the pace of change is speeding up. Today, corporate governance is the new watchword. Evidence of the new era is the number of companies that have appointed high-level officials to deal with the issues. Dealing with institutional shareholders is tricky. It's a disparate audience with many motives and interests, so no one approach will win all of them over. Large pension funds find it hard to sell stocks once they've bought them; either their investments are indexed, forcing them to hold the stocks in that index, or their holdings are so large that selling them would depress the price even further.

The objective for corporate governance is to find out what companies are doing and push them to do it better. As a result, corporations are growing more vigorous in managing their shareholder affairs. Increasingly, companies are redefining and elevating the shareholder relations role in an effort to influence the newly assertive stockholders. Through more open communication and active courting, companies seek to build stronger rapport with their owners and soften the sometimes adversarial relationship between the stockholder and management.

Investor Relations

In a general sense, it may be said that the principal purpose of the investor relations (IR) function, regardless of who performs it, is the enhancement of shareholder value. Executives of brokerage houses will acknowledge that a continuing investor relations program helps prepare the market for a public offering, and influences the credit ratings of fixed-income securities. Certainly, the importance of the stock price is not lost on those CEOs engaged in sell-offs, acquisitions, or other restructuring.

In today's environment companies must compete for investment capital--whether in the bond market or the equity market. But to secure recognition, the story of the enterprise must be told. Just as the advantages and uses of a company's service or product must be described and marketed, so also information about the value of its entity's securities and its financial prospects must be disseminated, understood, and accepted.

These changes also induced changes in the nature of the investor relations (IR) specialist. While communicators (public relations) and security analysts have predominated, gradually more and more individuals well acquainted with finance assume a greater role. The customers are far more sophisticated about what they needed. The IR function more and more evolved into a combination of two disciplines: communications and finance. Accordingly, in the 1990s, the investor relations function must not only communicate effectively about past performance, but it must more closely align itself with the strategic plans of management and tell its customers more about corporate goals and the entity's strengths and weaknesses.

Communication Vehicles for Investor Relations

The methods used to communicate investor-related messages, in no special order, include:

- Annual Report. The Security Exchange Commission (SEC)has authority to regulate external financial reporting by publicly traded companies. Nevertheless, its traditional role has been to promote disclosure rather than to exercise its power to establish accounting standards. To promote disclosure, the SEC has adopted a system that integrates the information required to be presented in annual reports to shareholders and in the SEC filings.
- Annual meeting with shareholders in which the financial and operating results are presented. The purpose of the annual meeting is to provide stockholders an opportunity to exercise their rights at annual or special meetings. The purpose is to elect new directors and to conduct necessary business.

- Quarterly reports to shareholders and the financial community regarding the financial and operational progress of the company in achieving the annual goals.
- Reports to the SEC. The purpose of the Securities Act of 1933 is to provide complete and fair disclosure to potential investors. The Securities Exchange Act of 1934 is intended to regulate trading of securities after initial issuance, provide adequate information to investors, and prevent insiders from unfairly using nonpublic information.
- Reports made to the SEC include:

  Various registration reports. Official reports (S series) containing a history of business and management and important financial information must be filed before security may be publicly offered.

  8-K Report of unscheduled material events of importance to shareholders or the SEC.

  10-K Official annual business and financial report filed with the SEC.

  11-K Annual report for employee stock purchase or similar plans.

  18-K Annual reports of foreign governments or political subdivisions.

  19-K Annual report form for issuers of American certificates.

  20-K Annual report form for foreign private issuers.

  7-Q Quarterly reports of real estate investment trusts or of companies whose major business is holding real estate.

  10-Q Quarterly reports containing specified financial information for each of the first three quarters of the year.
- Regular or special meetings with security analysts, institutional investors, brokers, and large individual investors often arranged in cooperation with one of the several associations or societies for analysts.
- Institutional advertising in newspapers or periodicals of a financial or general interest.
- Corporate announcements of special interest to investors or potential investors:
  a. New products or services
  b. Management changes
  c. Acquisitions and/or investment
  d. Reorganization attempts, etc: restructuring, unfriendly takeovers
- Videocassettes dealing with financial matters.
- Use of toll-free telephone numbers.
- Individual meetings with government representatives and the stock exchanges concerned with financial matters (IRS, SEC, etc).

There are three broad groups that the IR activity is primarily and continuously directed toward: (a) security analysts, (b) stockbrokers, and (c) large institutional investors. A most important source of information for the security analyst is management presentations. These are generally made to large institutional investors, brokers, or investment bankers, as well as the analysts themselves. The information gained from such meetings, plus annual and quarterly reports, form 10-Ks, etc., together with analysts' discussions, and articles about the company and the industry, enables the analyst to reach certain conclusions about the entity, and helps them predict financial performance.

Among other objectives, a purpose of the IR function is to enable the company to raise funds to meet its needs, on an acceptable economic basis, to enhance the long-term interests of the shareholders. The financial environment changes as the business cycle changes, or the perceived relative status of the company or industry changes. Indeed, even as the mood of the investor changes.

Recent developments include:

- Demands by some pension funds that they have a greater voice in certain company policy decisions.
- Increased agitation by unhappy shareholders about exorbitant levels of executive compensation or perquisites.
- Pressures by some institutional investors to make the board of directors more independent of the CEO, and
- Proposals by company management for the protection of shareholders in the event of an unsolicited bid for the corporation and vastly increased volatility in the stock market.

While circumstances will differ company by company, there will be instances where the CFO, probably assisted by the controller, will find it necessary to become more aggressive in cultivating the financial market, and take these actions: (1) establish specific financial market related objectives, which will be in the shareholders' interest, and (2) develop some methods of helping to reach these new objectives.

Keep the lines of communication open is rule number one, say investor relations specialists in dealing with an analyst who disagrees. Given that sell-side analysts, who after all work for brokers whose business is to sell stock, are prone by nature to praise. A hostile or critical individual can be as irritating as a persistent mosquito, especially for a company that is generally well liked. Whether he or she is poking holes in the company's predictions, performance, or image, it's hard to react calmly. There is a tendency to lash out, to stonewall, and to smash the pesky critter. Or, going to the other extreme, becoming obsessed by the criticism, determined at any cost to win the critic over.

It's easy to misunderstand the reasons the company is not getting the "buy" recommendation they are hoping for. One major communications problem stems from the fact that, although they don't usually say so, analysts sometimes aren't really analyzing the company's prospects as much as they are the outlook for its stock price. Sometimes when a company has a downturn in earnings, it goes into excruciating details as to why it incurred the loss but doesn't add what it's doing to fix matters. To an investor, what you're doing to fix matters is as material as what went wrong.

Econometrics

11

One of the more daunting tasks for companies in the petroleum industry is to try to anticipate which direction oil and gas prices are going to go. Price estimates provide one of the cornerstones in planning and analysis. Supply, demand, and price are constantly influenced by producers, processors, traders, and consumers making decisions they deem to be in their best economic interests. Price is both a product of this process and a key factor in the outcome. Price determines whether the consumers consume and whether producers produce.

FUNDAMENTALS OF SUPPLY AND DEMAND

The fundamentals of supply and demand are the birthplace of price forecasting. And while some economists dedicate a career to these complex issues, a foundation is provided here to introduce the concepts and drivers that influence prices.

Some published sources indicate the sensitivity of prices to disruptions in supply or to overproduction of, say, a million barrels of oil per day would impact oil prices by as much as $5/barrel or so. Other sources anticipate a more modest response. For example, some analysts would expect oil prices to drop by around $1.50 per barrel if oil production were to increase by about 1 million barrels per day

Certainly, chronic oversupply could cause continued downward pressure on prices. However, a million barrels per day of oversupply can represent less than a year of demand growth. Furthermore, the effect of over/undersupply is partly a function of timing. If a country or OPEC increased production by an extra million barrels per day just at the beginning of the winter heating season, then the price response might not be so dramatic. This would be particularly true if the oversupply was kept secret as opposed to announced. In November of 1997, Saudi Arabia announced it would add 2 million barrels oil per day (BOPD) to its existing 8 million BOPD quota. Prices began to respond immediately.

Demand Growth

Oil and gas represent roughly 62% of world energy consumption. Oil's share is 40% and these ratios are expected to stay about the same for quite a few years. Worldwide production in 1995 amounted to 69 million BOPD and 203 billion cubic feet of natural gas per day. Further statistics regarding world petroleum industry are found in Appendix 2.

In addition to the dynamic of over/undersupply is the rate of demand growth both anticipated and real. Overproduction of 1 million BOPD in the short-term can become the source for demand growth in the medium- to long-term. Furthermore, there are seasonal cycles such as the winter heating season which can increase short-term demand by as much as three million BOPD.

World oil demand growth averaged 7.5% per year from 1950 (and before) until 1972. Following the first embargo in 1973, demand growth stalled and declined during the early 1980s. From 1973 to 1996 annual demand growth averaged only 0.4%. Demand and real price growth trends from 1950 are summarized in Appendix 3.

It is often said that demand growth is expected to lie somewhere between world Gross Domestic Product (GDP) growth and population growth, although energy demand growth has exceeded GDP growth in non-OECD countries. But this is not expected to continue.

World Population Growth

World population in July 1995 was 5.9 billion. The population growth rate is on the order of 1.7%. This is the result of an average annual world birth rate of 26/1,000 population. World average death rate is 9/1,000 per year. Net growth then is 17/1,000 which amounts to about 99 million additional inhabitants per year on this planet. One unfortunate statistic associated with the annual 26/1,000 population growth rate which amounts to about five births per second is that 97% of those born are born into third-world conditions.

World Gross Domestic Product Growth

The gross domestic product worldwide (Gross World Product) was around $35 trillion in 1997. World economic growth averaged around 1.8% during the first five years of the 1990s. But the projected world GDP growth rate according to conventional wisdom is on the order of 3.5%. Most of this robust growth is expected from developing countries that hold 83% of the world's population. Industrial countries are expected to grow at an aggregate rate of around 2.8%. Conventional wisdom regarding the robust state of the industry is captured here with perspectives from the International Energy Agency (IEA), the Energy Information Administration (EIA), and Hossein Razavi, head of the Energy Division of the World Bank. (Figure 11–1)

World Oil Demand (MMBOPD)

	Year 2000	Year 2010
International Energy Agency World Energy Outlook Capacity Constraints Case Energy Savings Case	 77 75.7	 92 95.2
World GDP Growth 1992–2010		3.1%

Energy Information Administration
International Energy Outlook 1995

	1994	1995	2000	2005	2010	2015
World Oil Price Projections 2.4% annual growth rate to 2015	$15.52	$16.81	$19.27	$21.86	$23.70	$25.43

World Oil Demand Growth between 1990 and 2010: 1.5%

Hossein Razavi, World Bank

World Oil Consumption expected to grow annually: 1.7%
World Oil Consumption in 2010: 95 Million BOPD

The Asian Factor
(before the 1997/1998 currency crises)

Asia alone accounts for: 47% of Power Generation Growth
48% of increased Oil Demand
31% of increased Gas Demand

Figure 11–1 Crude Oil Demand Predictions (circa 1995)

Gas vs. Oil

The difference between oil and gas is fairly dramatic particularly in regions where established infrastructure and gas markets do not exist. The most distinguishing characteristic between oil and gas is the transportation function. The relative cost to transport gas compared to oil over large distances can be huge.

Demand growth for gas has centered on the stationary fuels market where it has outstripped demand growth for oil about five to one. Demand growth for oil is and has been primarily for transportation fuels. Gas markets have benefitted from air quality concerns and as a clean burning fuel, it has huge advantages. Furthermore, the growth of combined cycle power generation and its close relative, co-generation has created a booming market for gas in regions where many gas discoveries had little hope of development. The boom in gas-fired power generation was launched in the early 1990s when efficiency of converting gas to electric energy began to approach 50% with newer technology.

While oil is still a more valuable commodity than gas in nearly every setting, the focus has changed. Some companies are exploring for gas and as an example, it is often said, "Gas is coming of age in southeast Asia." The time will come when the market and infrastructure for gas in regions like southeast Asia will rival that of the Gulf of Mexico and Europe. It just takes time.

Nevertheless, nearly 10 billion cubic feet of gas a day is flared worldwide. Most of the gas flared in the world is gas associated with oil production. However, some flared gas is residue from liquid petroleum gas extraction facilities. Nigeria alone accounts for nearly a quarter of the world's flared gas. The alternative to flaring gas in many cases is reinjection, which can be prohibitively expensive or shut-in production. It is unfortunate to see so much gas flared, but the alternatives are even less pleasant from a financial point of view.

Supply/demand econometrics provide insight into the implications of oversupply or supply disruptions. In early 1997, Iraq began exporting oil in ever greater quantities. The anticipated price response to an extra 2 million barrels of oil per day from Iraq without any reduction in output from OPEC was certainly a downward price response. But how much? According to most sources the response of world oil prices would be on the order of $5 to $8 per barrel.

The Saudi production of around 8 million BOPD has been the only real swing production to accommodate oversupply from OPEC members or from non-OPEC producers. However, if the price response for an extra million barrels per day is only on the order of say $2.50 per barrel, then the Saudis have little incentive to scale back production. For example, with oil prices of $18 per barrel or so, Saudi's annual revenues would be on the order of $52.6 billion per year. If they scaled back production by 1 million BOPD and received the resulting higher price of around $20.50 per barrel (all things being equal), their revenues would be $52.4 billion. Not much difference, except the Saudis would lose market share to the tune of 1 million BOPD.

If the price/demand curve were steeper, then this formula would change in favor of cutting back to some degree. And, many believe the relationship is more dramatic than only $2.50 per barrel or so for every million BOPD of extra production.

ORGANIZATION OF PETROLEUM EXPORTING COUNTRIES (OPEC)

The demise of OPEC has been forecast on numerous occasions yet in the opinion of many the organization has managed to hold some reign on price fluctuations. Cheating is a fact of life among most members, but not the largest members.

The long-term dynamics include the demand side response to changes in oil prices. For every dollar above $20 per barrel, additional frontier provinces that otherwise hovered on the margin become viable. OPEC is still experiencing the effects of what happens when oil prices are too high. North Sea production in excess of 5 million BOPD in the mid-1990s was never anticipated. The UK and Norway producing all-out has been a thorn for OPEC, and it is unlikely OPEC would like to see any more provinces capture such a large share of the expected demand growth.

It is already anticipated that Africa, particularly west Africa, and South America, particularly Venezuela and Colombia, will provide the oil for much of the world demand growth into the early 2000s. Beyond that, it is currently expected that demand growth will be met by OPEC. But what about the former Soviet Union? With potential of 5 to 8 million BOPD of additional capacity, the former Soviet Union could be a big problem for OPEC and world oil prices. However, up to the mid 1990s the former Soviet Union has not managed to move very quickly toward realizing that potential. With world oil prices below $20 per barrel, it will take longer than if prices were at say $24 per barrel. The OPEC producers know this too.

The OPEC nations control three-fourths of the world's petroleum reserves, yet 85% of the OPEC reserves have no value from a present value point of view. This is one reason for the kind of behavior exhibited by the lead OPEC producers. These countries are faced with some heavy boundary conditions.

OPEC CAPACITY UTILIZATION

Some of the more bullish perspectives on oil prices come from the analysis of OPEC capacity utilization and the time and cost requirements to meet demand growth. During the mid 1990s there were a number of analysts that viewed OPEC's production at around 4% short of capacity. They explained that, with OPEC at 96% capacity, the potential for high oil prices in the near term are relatively certain. A near term squeeze if the OPEC producers do not gear up and make the necessary expenditures is just a matter of time. Demand growth of 1.5% equates to nearly 1 million BOPD per year. If OPEC is going to add capacity of around 1 million BOPD per year, the capital cost requirements will be on the order of $5 billion per year or so. This amounts to $5,000 per daily barrel of capacity which equals around $1.00 to $1.50 per barrel capital costs. North Sea reserves by comparison cost on the order of $10,000 to 12,000 per daily barrel. To add over 20 million BOPD capacity over the next 15 years then would require around $100 billion of capital.

In addition to these large expenditures, there is the consideration that it takes time to build the infrastructure to handle the added capacity. We may already be behind according to some.

	(MBOPD)			
	1988		1995[1]	
Country	Quota	Production	Quota	Production
Algeria	667	641	750	764
Ecuador[2]	221	305		
Gabon[3]	159	175	287	360
Indonesia	1,190	1,138	1,330	1,329
Iran	2,369	2,240	3,600	3,612
Iraq	1,540	2,688	400	600
Kuwait	996	1,335	2,000	1,785
Libya	996	1,025	1,390	1,375
Neutral Zone	[4]	313	[4]	430
Nigeria	1,301	1,390	1,865	1,887
Qatar	299	346	378	442
Saudi Arabia	4,343	4,946	8,000	7,852
UAE	948	1,474	2,161	2,204
Venezuela	1,571	1,649	2,359	2,609
Total OPEC	16,600	19,665	24,520	25,249
Non OPEC		43,530		42,051
Total World		63,195[5]		67,300[5]

[1] Quotas did not apply for the full year
[2] Ecuador dropped out at the end of 1992
[3] Gabon dropped out at the end of 1996
[4] Quotas do not apply, production shared by Kuwait and Saudi Arabia
[5] Rough estimate due to inclusion of gas liquids in world totals

Table 11–1 OPEC Production and Quotas

OPEC spare capacity was probably at an all-time low in the early 1990s particularly with the invasion of Kuwait by Iraq. In 1991, OPEC capacity was around 24.5 million barrels oil per day with only around 1 million BOPD spare capacity. Excess capacity in 1995 was estimated at around 3.3 million BOPD.

It is important to consider both the demand and supply side response to various oil prices. It has already been demonstrated that consumption can be dramatically curtailed with excessive prices. Figure 11-2 illustrates the balancing and countering effect of both supply and demand side responses. Figure 11-3 is a presentation of seasonal variations of supply.

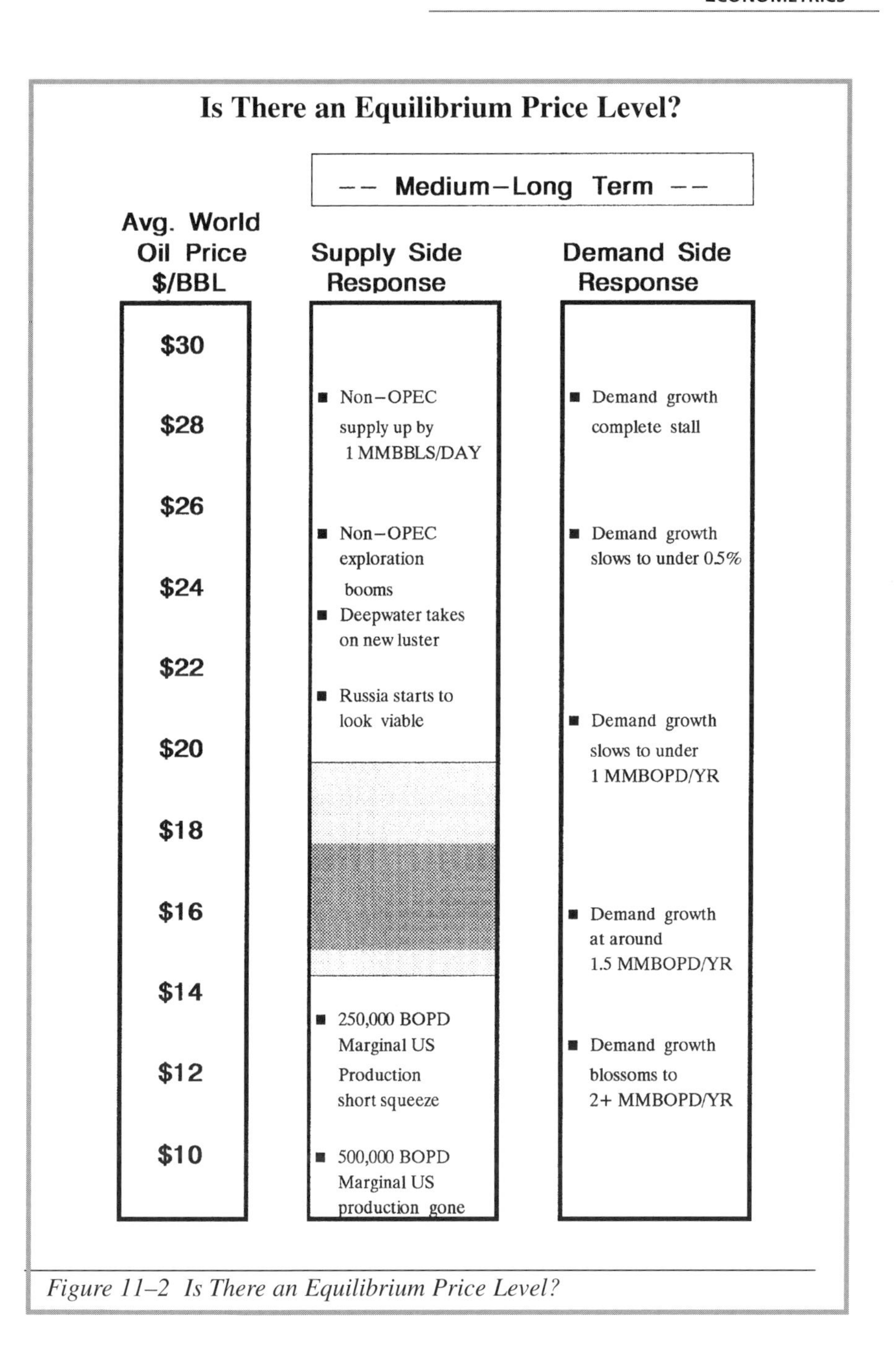

Figure 11–2 Is There an Equilibrium Price Level?

World Oil Supply/Demand — Seasonal Variations

	1995		1996		
Quarter	3rd	4th	1st	2nd	3rd
Supply	70.17	70.46	71.17	71.34	71.54
Demand	69.00	72.43	73.25	69.99	70.52

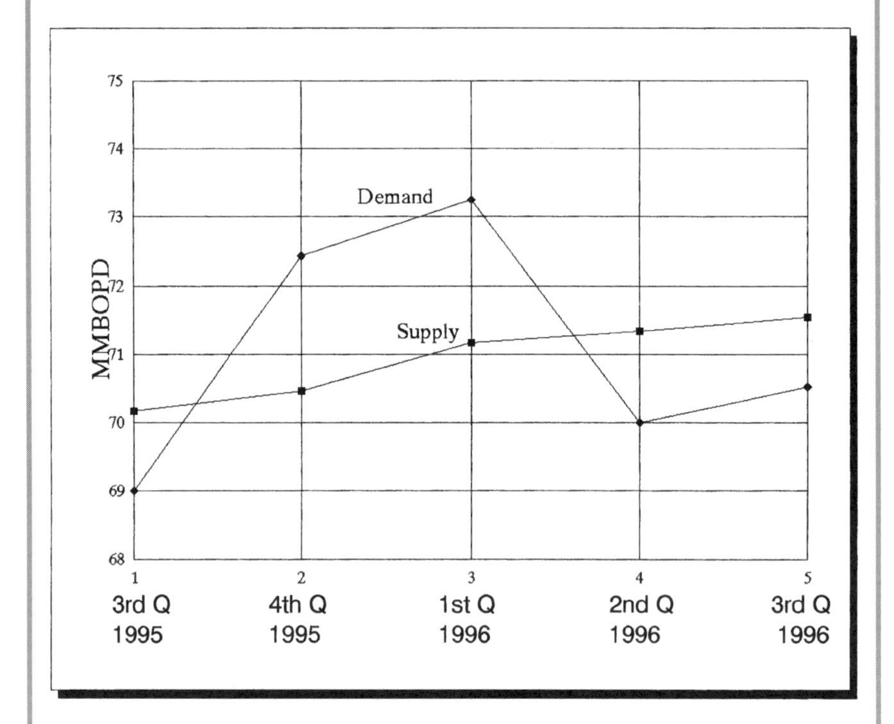

Demand includes crude oil, lease condensates, natural gas plant liquids, other hydrogen and hydrocarbons for refinery feedstocks, refinery gains, alcohol, and liquids produced from nonconventional sources.

Figure 11–3 World Oil Supply/Demand--Seasonal Variation

Appendix 1

Abbreviations and Acronyms

AFE	Authorization for Expenditure
API	American Petroleum Institute
B/CD	Barrels per Calendar Day (refinery, 365 days)
B/SD	Barrels per Stream Day (usually 330 days)
BBL	Barrel (crude or condensate), 42 U.S. gallons
BCF	Billion Cubic Feet of gas
BCPD	Barrels of Condensate Per Day
BOE	Barrels of Oil Equivalent, (see COE)
BOPD	Barrels of Oil Per Day
BOEPD	Barrels of Oil Equivalent Per Day
BWPD	Barrels of Water Per Day
BTU	British Thermal Unit
CAPM	Capital Asset Pricing Model
CEO	Chief Executive Officer
CFFO	Cash Flow From Operations
CIF	Cost, Insurance and Freight
COE	Crude Oil Equivalent (also called BOE)
COPAS	Council of Petroleum Accounts Societies
DB	Declining Balance
DCF	Discounted Cash Flow
DCFM	Discounted Cash Flow Method (not common)
DDB	Double Declining Balance Method
DD&A	Depreciation, Depletion and Amortization
DFL	Degree of Financial Leverage
DOL	Degree of Operating Leverage
DSO	Days' Sales Outstanding
DTL	Degree of Total Leverage
E&P	Exploration and Production
EBITXD	Earnings Before Interest, Taxes, Exploration Expenses and Depreciation
EBO	Equivalent Barrels of Oil (see BOE and COE)
EFT	Electronic Funds Transfer
EIA	Energy Information Administration
EME	Emerging Market Economies
EMV	Expected Monetary Value
EOQ	Economic Order Quantity
EOR	Enhanced Oil Recovery
EPS	Earnings Per Share
EUR	Estimated Ultimate Recovery
FASB	Financial Accounting Standards Board

FC	Full Cost Accounting
FIFO	First In, First Out
FMV	Fair Market Value
FOB	Free On Board
GAAP	Generally Accepted Accounting Principles
GAO	General Accounting Office
G&A	General and Administrative Expenses
G&G	Geological and Geophysical
GDP	Gross Domestic Product
GNP	Gross National Product
GOR	Gas Oil Ratio (usually cubic feet/barrel)
GWP	Gross World Product
GRR	Growth Rate of Return
IBRD	International Bank for Reconstruction and Development
IDA	International Development Association
IDC	Intangible Drilling Costs
IMF	International Monetary Fund
IRR	Internal Rate of Return
ISO	International Standards Organization
ITC	Investment Tax Credit
L/C	Letter of Credit
LDC	Local Distribution Company
LIBOR	London Interbank Offered Rate
LIFO	Last In, First Out
LNG	Liquified Natural Gas
LP	Limited Partnership
Ltd	Limited Liability (British Corporation)
LPG	Liquid Petroleum Gas
MBBLS	Thousand Barrels
MBO	Management Buy Out
MCFD	Thousand Cubic Feet per Day
MCP	Management Control Process
MCS	Monte Carlo Simulation
MER	Maximum Efficient Rate
MIGA	Multi-Lateral Investment Guarantee Agency
MIRR	Modified Internal Rate of Return
MMBBLS	Million Barrels
MMBPD	Million Barrels per Day
MMCF	Million Cubic Feet
MNOC	Multi-National Oil Company
NGL	Natural Gas Liquids
NPV	Net Present Value
NRI	Net Revenue Interest
OECD	Organization of Economic Co-operation and Development

OPEC	Organization of Petroleum Exporting Countries
ORI	Overriding Royalty Interest
ORRI	Overriding Royalty Interest
P10	Probability 10%
P50	Probability 50%
P90	Probability 90%
PCO	Parent Company Operations
PE	Price Earnings Ratio
PLC	(British) Public Limited Company
P/I	Profit to Investment Ratio
PV	Present Value
PVP	Present Value Profit
PVP/I	Present Value Profit to Investment Ratio
RDB	Regional Development Banks
RI	Royalty Interests
RLI	Reserve Life Index
ROA	Return on Assets
ROC	Return on Capital
ROE	Return on Equity
ROI	Return on Investment
RRA	Reserve Recognition Accounting
SCF	Statement of Cash Flows
SCFP	Statement of Changes in Financial Position
SE	Successful Efforts Accounting
SEC	Securities and Exchange Commission
SF	Swiss Francs
SLD	Straight-Line Decline
SMOG	Standard Measure of Oil and Gas
SYD	Sum-of-the-Years Digits
TCF	Trillion Cubic Feet
TCM	Total Cost Management
TLCB	Tax Loss Carry Back
TLCF	Tax Loss Carry Forward
VAT	Value Added Tax
WI	Working Interest
WPT	Windfall Profits Tax
$/BBL	Dollars per Barrel
$/BOE	Dollars per Barrel of Oil Equivalent
$/BOPD	Dollars per Barrel of Oil per Day
$/MCF	Dollars per Thousand Cubic Feet
$/MCFD	Dollars per Thousand Cubic Feet per Day

Appendix 2 Petroleum Industry Vital Statistics - 1995

Vital Statistics – 1995

World Average Export Price $16.78 /BBL — Avg. U.S. Oil Price $14.62; Avg. U.S. Gas Price $1.59

PRODUCTION

World Total Oil	69	MMBOPD [Demand growth 1–1.6%/year]
World Total Gas	203	BCFD
World Total BOE	103	MMBOE/D (6:1) [62% of total world energy consumption]
	89	MMBOE/D (10:1)
World Gas Flaring	10	BCFD

RESERVES

OIL	Total OPEC	778	Billion Barrels [85% has no NPV]
	Total World	1,007	Billion Barrels "Conventional Reserves"
GAS	Total CIS	1,977	TCF
	Total OPEC	2,037	TCF
	Total World	4,934	TCF
BOE	Total World	1,829	Billion BOE (6:1)
		1,500	Billion BOE (10:1)

US Total Drilling Rigs	723
World Total Drilling Rigs	1,711

Rigs Outside N. America **Land**	548
Offshore	210
Total rigs outside N. America	758

US Producing Gas Wells	294 Thousand	Avg. 200 MCFD
US Producing Oil Wells	582 Thousand	Avg. 11 BOPD
World Producing Oil Wells	904 Thousand	Avg. 68 BOPD
US Total Refineries	169	90,000 BOPD (avg) @ 92% capacity
World Total Refineries	706	104,000 BOPD (avg) @ 83% capacity

Appendix 3 Demand/Real Price Trends from 1950

Demand/Real Price

Year	World Crude MBOPD	Growth Rate Annual	Growth Rate 5-yr Avg.			U.S. Wellhead Prices Nominal $/BBL	U.S. Wellhead Prices Real 1995 $/BBL	PPI Growth Rate
1950	10,418	12%				$2.51	$11.45	3.9%
1951	11,734	13%				$2.53	$10.37	11.4%
1952	12,346	5%				$2.53	$10.66	−2.8%
1953	13,146	6%				$2.68	$11.46	−1.4%
1954	13,748	5%				$2.78	$11.85	0.3%
1955	15,414	12%	9.6%			$2.77	$11.77	0.3%
1956	16,734	9%				$2.79	$11.49	3.2%
1957	17,640	5%				$3.09	$12.37	2.9%
1958	18,104	3%				$3.01	$11.89	1.3%
1959	19,544	8%				$2.90	$11.43	0.2%
1960	20,973	7%	7.2%			$2.88	$11.34	0.1%
1961	22,382	7%		7.5%		$2.89	$11.42	−0.3%
1962	24,305	9%				$2.90	$11.43	0.2%
1963	26,092	7%				$2.89	$11.42	−0.2%
1964	28,103	8%				$2.88	$11.36	0.2%
1965	30,232	8%	8.8%			$2.86	$11.06	2.0%
1966	32,880	9%				$2.88	$10.79	3.2%
1967	35,256	7%				$2.92	$10.91	0.3%
1968	38,473	9%				$2.94	$10.71	2.5%
1969	41,610	8%				$3.09	$10.83	4.0%
1970	45,697	10%	10.2%			$3.18	$10.75	3.7%
1971	48,348	6%				$3.39	$11.10	3.3%
1972	50,823	5%				$3.39	$10.63	4.4%
1973	55,825	10%				$3.89	$10.78	13.2%
1974	55,985	0%			4.0%	$6.87	$16.03	18.8%
1975	52,960	−5%	3.2%			$7.67	$16.38	9.2%
1976	57,369	8%				$8.19	$16.71	4.6%
1977	59,595	4%				$8.57	$16.48	6.1%
1978	59,917	1%				$9.00	$16.05	7.8%
1979	62,375	4%				$12.64	$20.03	12.6%
1980	59,316	−5%	2.4%			$21.59	$29.99	14.1%
1981	55,748	−6%				$31.77	$40.44	9.1%
1982	53,475	−4%				$28.52	$35.57	2.1%
1983	53,232	−0%				$26.19	$32.27	1.2%
1984	54,496	2%				$25.88	$31.14	2.4%
1985	53,982	−1%				$24.09	$29.14	−0.5%
1986	56,227	4%		0.4%		$12.51	$15.58	−2.9%
1987	56,684	0%				$15.40	$18.69	2.6%
1988	58,745	1%				$12.57	$14.66	4.0%
1989	59,877	4%				$15.87	$17.64	5.0%
1990	60,594	2%	2.2%			$19.98	$21.43	3.6%
1991	60,219	1%				$16.53	$17.70	0.2%
1992	60,220	−1%				$16.00	$17.03	0.6%
1993	60,282	0%				$14.24	$14.94	1.5%
1994	61,326	0%				$13.19	$13.66	1.3%
1995	62,160	2%	0.5%			$14.62	$14.62	3.6%

Sources: Crude production – 20th Century Petroleum Statistics (1996) DeGolyer & MacNaughton
PPI all Commodities – Bureau of Labor Statistics Prices – Energy Statistics Sourcebook (1996) PennWell

Appendix 4 Average Wellhead Oil & Gas Prices—Total U.S. 1935–1995

YEAR	OIL $/BBL	GAS $/MCF	YEAR	OIL $/BBL	GAS $/MCF
1935	0.97	.06	**1965**	2.86	.16
1936	1.09	.06	**1966**	2.88	.16
1937	1.18	.05	**1967**	2.92	.16
1938	1.13	.05	**1968**	2.94	.16
1939	1.02	.05	**1969**	3.09	.17
1940	1.02	.05	**1970**	3.18	.17
1941	1.14	.05	**1971**	3.39	.18
1942	1.19	.05	**1972**	3.39	.18
1943	1.20	.05	**1973**	3.89	.22
1944	1.21	.05	**1974**	6.87	.30
1945	1.22	.05	**1975**	7.67	.45
1946	1.41	.05	**1976**	8.19	.59
1947	1.93	.06	**1977**	8.57	.80
1948	2.60	.07	**1978**	9.00	.91
1949	2.54	.06	**1979**	12.64	1.81
1950	2.51	.07	**1980**	21.59	1.59
1951	2.53	.07	**1981**	31.77	1.98
1952	2.53	.08	**1982**	28.52	2.46
1953	2.68	.09	**1983**	26.19	2.59
1954	2.78	.10	**1984**	25.88	2.66
1955	2.77	.10	**1985**	24.09	2.51
1956	2.79	.11	**1986**	12.51	1.94
1957	3.09	.11	**1987**	15.40	1.67
1958	3.01	.12	**1988**	12.58	1.69
1959	2.90	.13	**1989**	15.86	1.69
1960	2.88	.14	**1990**	20.03	1.71
1961	2.89	.15	**1991**	16.54	1.64
1962	2.90	.16	**1992**	15.99	1.74
1963	2.89	.16	**1993**	14.25	2.04
1964	2.88	.15	**1994**	13.19	1.88
			1995	14.62	1.59

From: *Oil & Gas Journal* Energy Database

Appendix 5 Conversion Factors

BTU Equivalents: Oil, Gas, Coal and Electricity

One British Thermal Unit (BTU) is equal to the heat required to raise the temperature of one pound of water (approximately one pint) one degree Fahrenheit at or near its point of maximum density.

One barrel (42 gallons) of crude oil
- = 5,800,000 BTUs of energy
- = 5,614 cubic feet of natural gas
- = 0.22 tons of bituminous coal
- = 1,700 kw hours of electricity

One cubic foot of natural gas (dry)
- = 1,032 BTUs of energy
- = 0.000178 barrels of oil
- = 0.000040 tons of bituminous coal
- = 0.30 kw hours of electricity

One short ton (2,000 pounds) of bituminous coal
- = 26,200,000 BTUs of energy
- = 5.42 barrels of oil
- = 25,314 cubic feet of natural gas
- = 7,679 kw hours of electricity

One kilowatt (kw) hour of electricity
- = 3,412 BTUs of energy
- = 0.000588 barrels of oil
- = 3.306 cubic feet of natural gas
- = 0.00013 tons of bituminous coal

Metric Conversions

One metric ton of crude oil
- = 2,204 pounds
- = 7–7.5 barrels of oil

One cubic meter of natural gas = 35.314 cubic feet

One cubic meter of liquid = 6.2888 barrels

One liter of liquid = 1.057 quarts

Distance

1 foot = 0.305 meters
1 meter = 3.281 feet
1 statute mile = 1.609 kilometers = 0.868 nautical miles
1 nautical mile = 1.852 kilometers = 1.1515 statute miles

Area

1 square mile	= 640 acres	= 2.59 sq km	= 259.0 sq hectares
1 square km	= 0.368 miles	= 100 hectares	= 247.1 acres
1 acre	= 43560 sq ft	= 0.405 hectares	
1 hectare	= 2.471 acres		

Volume

1 cubic foot = 0.028317 cubic meters
1 cubic meter = 35.514667 cubic feet
1 cubic meter = 6.2898 barrels
1 U.S. gallon = 3.7854 liters
1 liter = 0.2642 U.S. gallons
1 barrel = 42 gallons = 158.99 liters

Weight

1 short ton	= 0.907185 metric tons	= 0.892857 long tons
	= 2000 pounds	
1 long ton	= 1.01605 metric tons	= 1.120 short tons
	= 2240 pounds	
1 metric ton	= 0.98421 long tons	= 1.10231 short tons
	= 2204.6 pounds	

Appendix 6 Present Value of One-Time Payment

Present Value of $1 = $1/(1+i)^{n-.5}$

(Midyear discounting)

Year (n)	5%	10%	15%	20%	25%	30%	35%
1	.976	.953	.933	.913	.894	.877	.861
2	.929	.867	.811	.761	.716	.765	.638
3	.885	.788	.705	.634	.572	.519	.472
4	.843	.717	.613	.528	.458	.339	.350
5	.803	.651	.533	.440	.366	.307	.259
6	.765	.592	.464	.367	.293	.236	.192
7	.728	.538	.403	.306	.234	.182	.142
8	.694	.489	.351	.255	.188	.140	.105
9	.661	.445	.305	.212	.150	.108	.078
10	.629	.404	.265	.177	.120	.083	.058
11	.599	.368	.231	.147	.096	.064	.043
12	.571	.334	.200	.123	.077	.049	.032
13	.543	.304	.174	.102	.061	.038	.023
14	.518	.276	.152	.085	.049	.029	.017
15	.493	.251	.132	.071	.039	.022	.013
16	.469	.228	.115	.059	.031	.017	.010
17	.447	.208	.100	.049	.025	.013	.007
18	.426	.189	.087	.041	.020	.010	.005
19	.406	.171	.075	.034	.016	.008	.004
20	.386	.156	.066	.029	.013	.006	.003
21	.368	.142	.057	.024	.010	.005	.002
22	.350	.129	.050	.020	.008	.004	.002
23	.334	.117	.043	.017	.007	.003	.001
24	.318	.106	.037	.014	.005	.002	.001
25	.303	.097	.033	.011	.004	.002	.001

Appendix 7 Present Value of an Annuity

Present Value an Annuity of $1

(Midyear discounting)

Year	5%	10%	15%	20%	25%	30%	35%
1	.976	.953	.933	.913	.894	.877	.861
2	1.905	1.820	1.743	1.674	1.610	1.552	1.498
3	2.790	2.608	2.448	2.308	2.182	2.071	1.970
4	3.634	3.325	3.062	2.836	2.640	2.470	2.320
5	4.436	3.976	3.595	3.276	3.007	2.777	2.579
6	5.201	4.568	4.058	3.643	3.300	3.013	2.771
7	5.929	5.106	4.462	3.949	3.534	3.195	2.913
8	6.623	5.595	4.812	4.203	3.722	3.335	3.019
9	7.283	6.040	5.117	4.416	3.872	3.442	3.097
10	7.912	6.444	5.382	4.593	3.992	3.525	3.155
11	8.512	6.812	5.613	4.740	4.088	3.589	3.197
12	9.082	7.146	5.813	4.863	4.165	3.637	3.229
13	9.626	7.450	5.987	4.965	4.226	3.675	3.253
14	10.143	7.726	6.139	5.051	4.275	3.704	3.270
15	10.636	7.977	6.271	5.122	4.315	3.726	3.283
16	11.105	8.206	6.385	5.181	4.346	3.743	3.292
17	11.552	8.413	6.485	5.230	4.371	3.757	3.299
18	11.978	8.602	6.572	5.271	4.392	3.767	3.305
19	12.384	8.773	6.647	5.306	4.408	3.775	3.309
20	12.770	8.929	6.712	5.334	4.421	3.781	3.311
21	13.138	9.071	6.769	5.358	4.431	3.785	3.314
22	13.488	9.200	6.819	5.378	4.439	3.789	3.315
23	13.822	9.317	6.862	5.395	4.446	3.791	3.316
24	14.139	9.423	6.899	5.408	4.451	3.794	3.317
25	14.442	9.520	6.932	5.420	4.455	3.795	3.318

Appendix 8 Definitions and Formulas

Book Value = Shareholders' Equity

Adjusted Earnings/Share = Per Share Value of Net Income (Earnings) "Adjusted" for extraordinary items

All Long-term Obligations = Long-term Debt + Preferred Stock + Other Long-term Liabilities (Does not include Current Liabilities)

Appraised Share Value = Weighted average share value from Valuation Analysis Worksheet

Assets/Employee = Total Assets ÷ the number of employees

Average BOPD/Oil Well = Company Average Daily Oil Production ÷ Net Oil Wells

Average MCFD/Gas Well = Company Average Daily Gas Production ÷ Net Gas Wells

Book Value = Total company assets less total liabilities (owner's equity)

Cash Flow Interest Coverage = (Cash Flow + Provision for Income Taxes + Interest Expense – Deferred Taxes) ÷ (Interest Expense + Preferred Dividends)

Cash Flow/Share = Net Income (Adjusted for extraordinary items)
+ DD&A
+ Other Non-cash Expenses
+ Deferred Taxes
+ Exploration expenses

Current Ratio = Current Assets ÷ Current Liabilities

Debt Adjusted P/CF = (Market Capitalization + Long-term Debt + Other Obligations) ÷ (Cash Flow + Interest Expense)
(Interest expense can be "tax adjusted" by applying either statutory or effective tax rates.)
(Exploration expenses can be "tax adjusted" by applying either statutory or effective tax rates.)

Interest Rate = Interest Expense ÷ Long-term Debt

Interest Rate = (Annual Interest Expense + Preferred Dividends) ÷ (All Long-term Obligations)

Long-term Debt = Book Value of Debt or Reported Long-term Debt
Market Capitalization = Total Shares x Market Value/Share
Market Value/Share = Trading value of common stock as of the effective date of analysis

P/E Ratio = Market Value per share ÷ Adjusted Earnings per share

P/CF Ratio = Market Value per share ÷ Cash Flow per share

Percent Debt = (All Long-term Obligations – Working Capital) ÷ Total Capitalization

Production BOEPD/Employee = Average Daily Production in Barrels of Oil Equivalent (BOE) ÷ by number of employees (Gas converted at 6 MCF = 1 BBL)

Reserve Life Index (Oil & Gas) = Proved Developed Reserves ÷ Annual Production Rate

Reserves MBOE/Employee = Total Booked Barrels of Oil Equivalent Reserves (Gas converted at 6:1) ÷ number of Employees

Reserves/Well Gas = Proved Developed Gas Reserves ÷ Net Gas Wells

Reserves/Well Oil = Proved Developed Oil Reserves ÷ Net Oil Wells

Return on Assets = Adjusted Net Income ÷ Reported Total Assets

Return on Equity = Adjusted Net Income ÷ Book Value

Revenues/Employee = Total Revenues ÷ the number of Employees

Total Capitalization = (Market Capitalization + Current Liabilities + All Long-Term Assets Appraised Value Obligations) ÷ Appraised Value of all Assets

Working Capital = Current Assets – Current Liabilities

Appendix 9 Natural Gas Products

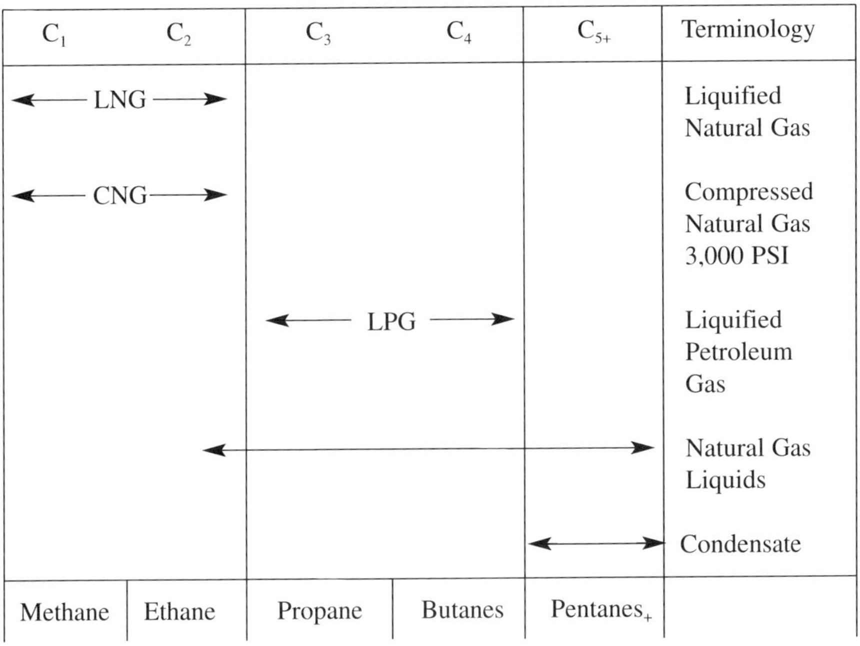

C_1	C_2	C_3	C_4	C_{5+}	Terminology
← LNG →					Liquified Natural Gas
← CNG →					Compressed Natural Gas 3,000 PSI
		← LPG →			Liquified Petroleum Gas
	←	—	—	→	Natural Gas Liquids
				←→	Condensate
Methane	Ethane	Propane	Butanes	$Pentanes_+$	

Liquified Natural Gas (LNG): –162°C

One MCF Gas = 43.57 pounds of LNG

One Ton of LNG = 46 MCF Gas

Appendix 10 Drilling Economics Algorithm

The basic equation for drilling cost analysis.

$$\$/\text{Foot} = (\$/\text{Bit} + \$/\text{Hour}\,(Tr + Td)) \div \text{Footage}$$

where

$\$/\text{Bit}$ = Cost of drill bit
$\$/\text{Hour}$ = Rig cost in \$/Hour
Tt = Trip time in hours
Td = Drilling time in hours
Footage = Number of feet drilled with bit

Glossary

Abrogate. To officially abolish or repeal a treaty or contract through legislative authority or an authoritative act.

Accelerated Depreciation. Writing off an asset through depreciation or amortization at a rate that is faster than normal accounting straight line depreciation. There are a number of methods of accelerated depreciation, but they are usually characterized by higher rates of depreciation in the early years in the life of the asset rather than the latter years. Accelerated depreciation allows for lower tax rates in the early years. (see Depreciation)

Accounting Cycle. All steps in the accounting process including posting entries, recording and analyzing transactions, adjusting and closing accounts, and preparing financial statements.

Accounting Rate of Return. Project average after-tax net income divided by the average cost of the investment over its estimated life.

Accounts Receivable. Short-term (current) assets arising from sales on credit to customers.

Accrual Accounting. The recognition of an expense or revenue that has not been recorded. As opposed to "cash" accounting.

Ad Valorem Tax. Latin meaning *according to value*. A tax on goods or property based upon value rather than quantity or size.

Agency Problem. Conflict of goals between a firm's shareholders and its managers.

Amortization. An accounting convention designed to emulate the cost or expense associated with reduction in value of an intangible asset over a period of time. Amortization is to intangible expenditures what depreciation is to tangible expenditures. (see Depreciation) Amortization is a noncash expense. Similar to depreciation of tangible capital costs, there are several techniques for amortization of intangible capital costs:

Straight-Line Decline (SLD)
Double Declining Balance (DDB)
Declining Balance (DB)

Sum-of-Years-Digits (SYD)
Unit-of-Production

ARBITRATION. A process in which parties to a dispute agree to settle their differences by submitting their dispute to an independent individual or group for settlement. Typically each side of the dispute chooses an arbitrator and those two choose a third. The third arbitrator acts as the chairman of the tribunal which then hears and reviews both sides of the dispute. The tribunal then renders a decision that is final and binding.

ARBITRAGE. Action to capitalize on a discrepancy in quoted prices; in many cases, there is not investment of funds tied up for any length of time.

ASIAN DOLLAR MARKET. Market in Asia where banks collect deposits and make loans denominated in U.S. dollars.

ASK PRICE. Price at which a trader of foreign exchange (typically a bank) is willing to sell a particular currency.

AUDITING. The principal function of a certified public accountant (CPA); the process of examining and testing the financial statements of a company in order to render an independent opinion as to the fairness of their presentation.

AUDITOR'S REPORT OR ACCOUNTANT'S REPORT. A report by an independent public accountant or accounting firm that accompanies the financial statements and contains the accountant's opinion regarding the fairness of presentation of the financial statements.

BETA. Measure of the volatility of a stock relative to either stock market index or to an index based upon a universe of industry related stocks. If a stock's price tends to follow its industry group up or down in synchronization, the stock will have a Beta of one. Stocks that rise more sharply than the stock market or an industry related group in a bull market and fall more sharply in a bear market will have a Beta greater than one. A low Beta stock will exhibit a relatively stable performance during market fluctuations. For example, if every time the market went up 10%, the stock of Company X would only go up 7%. The Beta for the company relative to the market would be 70% (or .7). (see Capital Asset Pricing Model)

Bilateral Agencies. These are development institutions set up in industrialized countries to support the investment and technical assistance requirements of developing countries. These agencies normally receive part or all of their funds from their respective governments.

Bond Ratings. Quality ratings that reflect the probability of the firm issuing the bonds defaulting on the issue in question. Each bond issue, short- or long-term, is independently rated. The two major rating agencies are Standard and Poor's and Moody's Investors Service.

Book Value. (1) The value of the equity of a company. Book value of stock is equal to the equity divided by the number of shares of common stock. Fully diluted book value is equal to the equity less any amount that preferred shareholders are entitled to, divided by the number of shares of common stock. (2) Book Value of an asset or group of assets is equal to the initial cost less depreciation, depletion, and amortization (DD&A).

Bretton Woods Agreement. Conference held in Bretton Woods, New Hampshire, in 1944 resulting in agreement to maintain exchange rates of currencies within very narrow boundaries. This agreement lasted until 1971. The concept of the World Bank began at this conference. The World Bank's purpose was to rebuild affected nations after World War II.

Call Option. Grants the right to purchase a specific asset at a specific price (exchange rate) within a specific period of time.

Callable Bonds. Bonds that a corporation has the option before maturity of buying back and retiring at given price, usually above face value.

Callable Preferred Stock. Preferred stock that may be redeemed and retired by the corporation at its option.

Capital Asset Pricing Model (CAPM). Sophisticated model of the relationship between expected risk and expected return. The model is based upon expected value theory and the theory that investors demand higher returns for higher risks. The return on an asset or security should be equal to a risk-free return (such as from a short-term treasury security) plus a risk premium.

RRR = Rf + Bi(Rm – Rf)

RRR = Required Rate of Return

Rf = Risk Free Rate of Return (U.S. Treasuries)

Rm = Market Rate of Return

Bi = Beta of the investment

CAPITAL BUDGETING. The combined process of analyzing alternative investment options, preparing reports for management, selecting the best alternatives, and rationing available capital funds among competing opportunities.

CAPITAL EXPENDITURE. An expenditure for the purchase or expansion of physical assets, such as property, plant and equipment (PP&E).

CAPITAL GAIN. The difference between the purchase price of an asset and the selling price, if the asset sells for more than the purchase price.

CAPITAL LEASE. A long-term lease that is in effect an installment purchase of assets; recorded by entering on the books a corresponding liability at the present value of the lease payments; each lease payment is partly a repayment of debt and partly an interest payment on the debt.

CAPITALIZE. (1) In an accounting sense the periodic expensing (amortization) of capital costs such as through depreciation or depletion. (2) To convert an (anticipated) income stream to a present value by dividing by an interest rate, as in the dividend discount model. (3) To record capital outlays as additions to asset value rather than as expenses.

CAPITALIZATION. All money invested in a company including long term debt (bonds), equity capital (common and preferred stock), retained earnings and other surplus funds.

CAPITALIZATION RATE. The rate of interest used to convert a series of future payments into a single present value.

CASH ACCOUNTING. A basis of accounting under which revenues and expenses are accounted for on a cash received and cash paid basis.

CASH FLOW. (1) In a loose sense, defined as net income plus depreciation, depletion, and amortization and other noncash expenses. Usually synonymous with cash earnings and operating cash flow. (2) An analysis of all the changes that effect the cash account during an accounting period.

Cash Flow Forecast. A forecast or budget that shows the firm's projected ending cash balance and the cash position for each month of the year so that periods of high or low cash availability can be anticipated; also called a *cash budget*.

Cash Management. Strategies for optimization of cash flows and investment of excess cash.

Central Exchange Rate. Exchange rate established between two European currencies through the European Monetary System arrangement. The exchange rate between the two currencies is allowed to move within bands around that central exchange rate.

Centralized Cash Management. Consolidation of cash management decisions for all Multinational Oil Company units, usually at the parent's location.

CIF. Cost insurance and freight is included in the contract price for a commodity. The seller fulfills his obligations when he delivers the merchandise to the shipper, pays the freight, and insurance to the point of (buyer's) destination and sends the buyer the bill of lading, insurance policy, invoice, and receipt for payment of freight. The following example illustrates the difference between an FOB Jakarta price and a CIF Yokohama price for a ton of LNG. (see FOB).

FOB Jakarta	$170/ton	also called "netback price"
	+ 30/ton	Freight Charge
CIF Yokohama	$200/ton	

Closely Held Corporation. A corporation whose stock is owned by a few individuals and whose securities are not traded publicly.

Commercial Discovery. In popular usage the term applies to any discovery which would be economically feasible to develop under a given fiscal system. As a contractual term, it often applies to the requirement on the part of the contractor to demonstrate to the government that a discovery would be sufficiently profitable to develop from both the contractor and government's point of view. A field that satisfied these conditions would then be granted "commercial status" and the contractor would then have the right to develop the field.

COMMON STOCK EQUIVALENTS. Convertible securities that, at the time of issuance, have a value that is closely related to the conversion value of the stock into which they can be converted. Warrants, rights and options are always common stock equivalents.

COMPENSATING BALANCE. A minimum account balance that a bank requires a company to keep in its bank account as part of a credit granting arrangement.

COMPOSITE BOND RATING. Weighted average of all bond ratings in a particular grouping or industry, e.g., petroleum industry.

COMPLEX CAPITAL STRUCTURE. A capital structure that includes additional securities such as convertible preferred stocks or bonds that can be converted into common stock. (see Common Stock Equivalents).

CONSORTIUM. A group of companies operating jointly, usually in a partnership with one company as operator in a given permit, license, contract area, block, etc.

CONTRACTOR. An oil company operating in a country under a production sharing contract or a service contract on behalf of the host government for which it receives either a share of production or a fee.

CONTRACTOR TAKE. Total contractor after-tax share of profits.

CONTRIBUTED OR PAID IN CAPITAL. The portion of owners' equity representing amount of assets invested by stockholders.

CONVERTIBLE BONDS. Bonds that may be exchanged for other securities such as common stock.

COST CENTER. A segment of an organization (by region or activity) where costs are accounted for separately.

COST INSURANCE AND FREIGHT (CIF). A transportation term that reflects the price of a commodity at the point of sale which includes all transportation costs including insurance, etc. (see CIF).

COST METHOD. Method of accounting for long-term investments when the corporation has neither significant influence nor control of the investment. The investment is recorded at cost and dividends are recognized as income when they are received.

Cost of Capital. The minimum rate of return on capital required to compensate debt holders and equity investors for bearing risk. Cost of capital is computed by weighting the after-tax cost of debt and equity according to their relative proportions in the corporate capital structure.

Cost of Goods Sold. Item on income statement computed by subtracting merchandise inventory at the end of the accounting period from the goods available for sale. It is deducted from gross revenue to yield gross profit.

Creeping Nationalization or Creeping Expropriation. A subtle means of expropriation through expanding taxes, restrictive labor legislation, or labor strikes, withholding work permits, import restrictions, price controls and tariff policies. The difference between nationalization and expropriation is that nationalization is usually on an industrywide level and expropriation focuses on a particular company.

Cross Futures Analysis. A comparison of the lists of events leading to each scenario and using an analytical process called the five forces. The five forces effecting an endstate consist of new entrants, power of shippers, rivalry, substitutes, and power of buyers.

Cumulative Preferred Stock. Preferred stock on which unpaid dividends (if they exist) accumulate and must be paid before dividends are paid on common stock.

Currency Bid-Ask Spread. Difference between the price at which a bank is willing to buy a currency versus the price at which it will sell that currency.

Currency Bid Price. Price that a trader of foreign exchange (typically a bank) is willing to sell a particular currency.

Current Assets. Cash or other liquid assets that are reasonably expected to be realized or can be converted to cash within one year.

Current Liabilities. Debt and other obligations that are due within one year. Current liabilities usually include accounts payable, taxes, wage accruals, and the portion of long term-debt and notes payable within 12 months.

Current Ratio. The ratio of current assets to current liabilities. Related to working capital which is equal to current assets less current liabilities.

CURTAILED PRODUCTION. Oil or gas production that is producing at a relatively reduced rate due to market or regulatory restraints.

DEBENTURE. A financing agreement issued without collateral (unsecured).

DEBENTURE BONDS. Bonds issued without collateral (unsecured).

DEBT SERVICE. Cash required in a given period, usually one year, for payments of interest and current maturities of principal on outstanding debt. In corporate bond issues, the annual interest plus annual sinking fund payments.

DEBT-TO-EQUITY RATIO. Ordinarily a ratio that measures the relationship of the claims on assets of creditors relative to shareholders. (1) Total long-term debt divided by common shareholders' equity. This is a measure of financial leverage. (2) Total liabilities divided by total shareholders' equity. This shows to what extent owner's equity can cushion creditor's claims in the event of liquidation. (3) Long-term debt and preferred stock divided by common stock equity (Although some view preferred stock as a class of equity).

DEFEASANCE. Short for in-substance defeasance. A technique whereby a corporation discharges old, low-rate debt without repaying it prior to maturity. The corporation uses newly purchased securities with a lower face value, but paying higher interest or having a higher market value. The objective is a cleaner (more debt free) balance sheet and increased earnings in the amount by which the face amount of the old debt exceeds the cost of the new securities.

DEFERRED CHARGE OR DEFERRED COST. A payment that is carried forward (as an asset) and not recognized as an immediate expense. Examples would be prepaid expenses that provided services, or an insurance premium payment that provided coverage, beyond the accounting period in which the payments were made.

DEFERRED INCOME TAXES. The difference between the income tax expense and the current income taxes payable accounts.

DEPLETION. The proportional allocation of the cost of a natural resource to the units removed; (1) Economic depletion is the reduction in value of a wasting asset by the removal of minerals.

(2) Depletion for tax purposes (depletion allowance) deals with the reduction of taxable income from mineral resources due to exhaustion of the resource by production or mining.

DEPRECIATION (DEPRECIATION EXPENSE). An accounting convention designed to emulate the cost or expense associated with reduction in value of an asset due to wear and tear, deterioration or obsolescence over a period of time. Depreciation is a noncash expense. There are several techniques for depreciation of capital costs:

Straight Line Decline (SLD)
Declining Balance (DB) *
Sum of Year Digits (SYD) *
Double Declining Balance (DDB)**
Unit-of-Production (UOP) * (see Unit-of-Production Depreciation)

* Referred to as accelerated depreciation.
**The DDB method is the maximum rate allowable for income tax purposes in the U.S.

DILUTION. The effect on book value per share or earnings per share if it is assumed that all convertible securities (bonds and preferred stock) are converted and/or all warrants and stock options are exercised.

DISCOUNTED CASH FLOW. The process of discounting future cash flows back to the present using an anticipated discount rate.

DIVIDEND. A distribution of assets of a corporation to its shareholders.

DIVIDEND YIELD. A ratio that measures the current return to an investor in a stock—annual dividend rate divided by stock price.

EARLY EXTINGUISHMENT OF DEBT. The extraordinary gain that occurs when a company purchases its bonds on the open market and retires them rather than waiting to pay them off at face value.

EARNINGS. The amount of profit realized after deduction of all costs, expenses and taxes. Also referred to as net income and net earnings.

EARNINGS PER SHARE. Net income divided by the weighted average number of shares of common stock and common stock equivalents.

ELECTRONIC FUNDS TRANSFER (EFT). This is a procedure in which customers electronically transfer funds from their bank account to the bank account of the recipient. The objective is to reduce cash float and manual procedures.

EQUITY. Residual interest in an asset or assets after deducting liabilities. Also known as book value.

EQUITY METHOD. A method of accounting for long-term investments under which the investor records the initial investment at cost and records proportional share of earnings and dividends in the investment account.

EUROCURRENCY. A collection of banks that accept deposits and provide loans in large denominations and in a variety of currencies.

EURODOLLARS. U.S. dollars deposited in Europe.

EXCISE TAX. A tax applied to a specific commodity such as tobacco, coffee, gasoline, or oil based either on production, sale or consumption.

EX-DIVIDEND. Stock sold after the date of record set for distribution of a dividend—the right to the dividend does not transfer with the sale.

EXPENSE. (1) In a financial sense, a noncapital cost associated most often with operations or production. (2) In accounting, costs incurred in a given accounting period as expenses and charged against revenues. To expense a particular cost is to charge it against income during the accounting period in which it was spent. The opposite would be to capitalize the cost and charge it off through some depreciation method.

EXPLORATORY WELL. A well drilled in an unproved area. This can include: (1) a well in a proved area seeking a new reservoir in a significantly deeper horizon, (2) a well drilled substantially beyond the limits of existing production.

EXTRAORDINARY ITEMS. Nonrecurring, usually one-time events that require a separate income statement entry as well as explanation in the footnotes. These can include write-off of a segment, sale of a subsidiary, or negative impact of a legal decision.

FAIR MARKET VALUE OF RESERVES-IN-THE-GROUND. Often defined as a specific fraction of the present value of future net cash flow discounted usually at a specific discount rate. The most common usage defines FMV at two-thirds to three-fourths of the present value of future net cash flow discounted at the prime interest rate plus .75 to 1 percentage point.

Financial Accounting Standards Board (FASB). The body responsible for developing and issuing rules on accounting procedure; issues Statements of Financial Accounting Standards (SFAS).

Finding Cost. The amount of money spent per unit (barrel of oil or MCF of gas) to acquire reserves. May include discoveries, acquisitions and revisions to previous reserve estimates.

First In, First Out (FIFO). The method of inventory accounting where it is assumed that inventory is used or sold in the chronological order in which it was acquired. The formula is: Inventory at the beginning of period plus purchases during accounting period minus ending inventory equals costs of goods sold. In a period of rising prices the FIFO (first in, first out) method produces higher ending inventory, a lower cost of goods and a higher gross profit than the LIFO method.

Float. The difference between the total dollar amount of checks drawn on a bank account and the balance shown on the bank's books. Many strategies are used to "play the float":

- Concentration banking is a means of accelerating the flow of funds of a firm by establishing strategic collection centers. Instead of a single collection center located at central company headquarters, multiple collection centers are established. The purpose is to shorten the period between the time a customer mails in his payment and the time when the company has the use of the funds.
- Lockbox system is another means of accelerating the flow of funds. Remittances are received by a collection center and deposited in the bank after processing. The purpose of a lockbox arrangement is to eliminate the time between the receipt of remittances by the company and their deposit in the bank.
- Electronic Funds Transfer can accelerate receipt of funds when customers either use electronic funds for transfer from customer bank to the company's bank.
- Effective control of disbursements can also improve turnover of cash. Whereas the underlying objective of collections is maximum acceleration, the objective in disbursements is to slow them down as much as possible.

FOB (Free on Board). A transportation term that usually means that the invoice price includes transportation charges to a specific destination. Title is usually transferred to the buyer at the FOB point by way of a bill of lading. For example, FOB New York means the buyer must pay all transportation costs from New York to the buyers receiving point. FOB plus transportation costs equals CIF price. (see CIF, Cost Insurance Freight).

FOB DESTINATION. Supplier bears transportation costs to the destination.

FOB SHIPPING POINT. Buyer bears transportation costs from point of origin (shipping point).

FOREIGN TAX CREDIT. Taxes paid by a company in a foreign country may sometimes be treated as taxes paid in the company's home country. These are creditable against taxes and represent a direct dollar-for-dollar reduction in tax liability. This usually applies to foreign income taxes paid and credited against home country income taxes. Other taxes that may not qualify for a tax credit may nevertheless qualify as deductions against home country income tax calculation.

FULL COST ACCOUNTING. A method of accounting for oil and gas exploration and development costs under which the costs of successful and unsuccessful exploration are capitalized. (see Successful Efforts Accounting).

FULLY DILUTED EARNINGS PER SHARE. Net income applicable to common stock divided by the sum of the weighted-average common stock and common stock equivalents and other potentially dilutive securities. (see Common Stock Equivalents).

FUNCTIONAL CURRENCY. Currency of the country where a subsidiary carries on most of its business.

FUTURES. Contracts specifying a standard volume of a particular asset to be exchanged on a specific settlement date.

GENERALLY ACCEPTED ACCOUNTING PRINCIPLES (GAAP). The conventions, rules, and procedures that define accepted accounting practice at a given time. Known in some countries as Generally Accepted Accounting Practices.

GOLD PLATING. When a company or contractor makes unreasonably large expenditures due to lack of cost cutting incentives. This kind of behavior could be encouraged where a contractors compensation is based in part on the level of capital and operating expenditure.

GROSS DOMESTIC PRODUCT (GDP). Measures the total value of goods and services produced within a country over a given period, usually a year.

GROSS NATIONAL PRODUCT (GNP). Comprises GDP plus income to domestic entities from foreign holdings minus income to foreign entities from domestic holdings.

HURDLE RATE. Term used in investment analysis or capital budgeting that means the required rate of return in a discounted cash flow analysis. Projects to be considered viable must at least meet the hurdle rate. Investment theory dictates that the hurdle rate should be equal to or greater than the cost of capital.

INCOME FROM OPERATIONS (OPERATING INCOME). See Operating Profit.

INCONVERTIBILITY. Inability of a foreign contractor to convert payments received in soft local currency into home country or hard currency such as dollars, pounds, or yen.

INDEPENDENT OIL COMPANY. A company that is involved primarily in exploration and production (the upstream sector).

INDEPENDENT PRODUCER. A loose term that generally refers to an individual or a small company. The term usually implies that the independent is not integrated.

INFLATION. The increase in the general price of goods and services.

INTANGIBLE DRILLING AND DEVELOPMENT COSTS (IDCs). Expenditures for wages, transportation, fuel, and fungible supplies used in drilling and equipping wells for production.

INTANGIBLES OR INTANGIBLE ASSETS. All intangible assets such as goodwill, patents, trademarks, unamortized debt discounts and deferred charges. All long-term assets that have no physical substance.

INTEGRATED OIL COMPANY. A company that has operations downstream as well as upstream. The term usually implies exploration and production integrated with transportation, refining and marketing operations. Typically the term is used for nonmajor oil companies.

INTERNAL RATE OF RETURN (IRR). The discount rate that yields a present value of future cash flow from an investment equal to the cost of the investment.

INTERNATIONAL BANK FOR RECONSTRUCTION AND DEVELOPMENT (IBRD). Also referred to as the World Bank, was established in 1944 to enhance economic development by providing loans to countries.

INTERNATIONAL DEVELOPMENT ASSOCIATION (IDA). Established to stimulate country development; it was especially suited for less prosperous nations since it provided loans at low interest rates.

INTERNATIONAL FINANCE CORPORATION (IFC). Established to promote private enterprise within countries; it can provide loans to and purchase stock of corporations.

INTERNATIONAL MONETARY FUND (IMF). Oversees the economic policies of its member countries (rich and poor) and uses its financial resources to help them through periods of adjustment.

JUST-IN-TIME INVENTORIES. In an effort to minimize investment in inventories, companies make agreements with suppliers to cooperate in achieving this objective. Some of the strategies and agreements include:

- Vendors agree to maintain inventories at the customer's worksite and charge the customer when goods are consumed.
- Utilize computer technology to integrate purchasing, receive goods, maintain inventories, and pay vendor.
- Partnership with supplier to provide materials only as needed.

LAST IN, FIRST OUT (LIFO). An inventory accounting method that ties the cost of goods sold to the cost of the most recent purchases. The formula is: cost of goods sold equals beginning inventory plus purchases minus ending inventory. Balance sheet inventories during times of inflation are typically lower than market value of inventories under LIFO accounting. The difference is referred to as LIFO cushion. In contrast to First in, First Out (FIFO) during periods of rising prices, LIFO produces a higher cost of goods sold that results in lower gross profit and taxable income.

LETTER OF CREDIT. An instrument or document from a bank to another party indicating that a credit has been opened in that parties favor guaranteeing payment under certain contractual conditions. The conditions are based upon a contract between the two parties. Sometimes called a performance letter of credit which is issued to guarantee performance under the contract.

LETTER OF INTENT. A formal letter of agreement signed by all parties to negotiations after negotiations have been completed outlining the basic features of the agreement, but preliminary to formal contract signing.

LEVERAGE. The relationship of debt and equity, often measured by the debt-to-equity ratio defined by total-long term debt divided by shareholders equity. The greater the percentage of debt, the greater the financial leverage. For example, assume that:

1. Company X can earn 15% rate of return (ROR) after-tax on invested capital.
2. The company can borrow at 10% interest
3. The company borrows 50% of its capital
4. Total investment is $1,000

With a rate of return of 15% the company earns $150 on its investment.

The company pays $50 interest (10% on $500). The after-tax cost of the interest is $33 because interest is deductible.

$ 500	Invested capital
500	Borrowed capital
$1,000	Total capital

150	After-tax Rate of Return before interest expense
33	After-tax Cost of Debt
$ 177	After-tax Return with after-tax interest expense deducted

$ 177	After-tax leveraged return	= 23.4%
$ 500	Invested Equity Capital	

Return on Equity = 24.4%

With borrowed funds the return is lower at $177 instead of $150, but the rate of return goes from 15% to 23.4%. This is what is meant by financial leverage. Leverage can enhance profitability of invested capital, but it can also enhance risk. Just as leverage can magnify profitability, it can also magnify losses.

LIABILITY. A debt of the business—amount owed or obligation to perform a service. A claim on assets.

LIFTING. The amount of crude oil an operator produces and sells or the amount each working interest partner (or the government) takes. The liftings may actually be more or less than actual entitlements which are based on royalties, working interest percentages and a number of other factors. If an operator or partner has taken and sold more oil than it was actually entitled to, then it is in an "overlifted" position. Conversely, if a partner has not taken as much as it was entitled to, it is in an "underlifted" position. (See Nomination and Entitlements).

LIQUIDATE. The act of selling an asset or security for cash.

LIQUIDITY. The position of having enough funds on hand to pay bills when due.

LOCKBOX. A lockbox system is another means of accelerating the flow of funds. The purpose of a lockbox is to eliminate the time between the receipt of remittances by the company and their deposit in the bank. The arrangement usually is on a regional basis, with the company choosing regional banks according to its billing patterns. Before determining the regions to be used, a feasibility study is made of the availability of checks that would be deposited under alternative plans. The company rents a local post office box and authorizes its bank in each of these cities to pick up remittances in the box. The bank picks up the mail several times a day and deposits the checks in the company's account. Electronic transmission is sent to the company by the bank of total cash received and customers transmitting. Some companies use the electronic data file for application of cash to the accounts receivable system.

LONDON INTERBANK OFFERED RATE (LIBOR). The rate that the most creditworthy international banks that deal in Eurodollars will charge each other. Thus, LIBOR is sometimes referred to as the Eurodollar Rate. International lending is often based on LIBOR rates, for example a country may have a loan with interest pegged at LIBOR plus 1.5%.

LONG-TERM DEBT. Liabilities that are expected to fall due after 12 months. (Long-term liabilities).

LOWER-OF-COST-OR-MARKET (LCM) RULE. Method of accounting for assets such as inventory of investments in securities on the balance sheet at either cost or market value whichever is lower.

MAJOR OIL COMPANY. The term major refers to the largest integrated oil companies. These companies will often be fully integrated with exploration, production, transportation, refining, petrochemicals, and marketing operations.

Marginal Tax Rate. The tax rate that applies to the last increment of taxable income.

Market Capitalization. The market capitalization of a company is equal to the number of shares of common stock times the market price per share.

Market Risk. The volatility or fluctuation of the price of a stock in relation to the volatility or fluctuation of the prices of other stocks. (See Beta)

Marketable Securities. Investment in securities that are readily marketable.

Maximum Cash Impairment. The maximum amount of capital investment required for a project considering that the total capital investment may be greater, but may be offset by cash flow during the capital investment stage. Therefore, maximum cash impairment is usually less than the project total capital cost requirements. The maximum cash impairment figure is the answer to management's question; "How much money do we have to come up with to do this project?"

Monte Carlo Simulation. Method of numerical simulation using random number generation for statistically sampling possible variables. Requires numerous iterations of basic algorithm to yield hopefully a statistically significant sampling of input variables to create a distribution of possible results.

Mortgage. A type of long-term debt secured by real property that is paid in equal installments.

Multilateral Investment Guarantee Agency (MIGA). An affiliate of the World Bank to encourage foreign investment in developing countries by providing investment guarantees against the risks of currency transfer, expropriation, war, civil disturbances, and breach of contract by the host government.

Net Earnings or Net Income or Net Profit. see Earnings.

Net Revenue Interest. This is the representative oil and gas ownership after deducting royalty claims on the oil and gas production. A working interest holder who owns an 80% working interest in a property with a 10% royalty obligation has a 72% net revenue interest.

Net Transaction Exposure. Consideration of inflows and outflows in a given currency to determine the exposure after offsetting inflows against outflows.

NETTING. The process of combining cash receipts and payments to determine the net amount to be owed by one subsidiary to another subsidiary in the same local currency. Netting is used by foreign branches of oil and other firms to minimize exposure to exchange rate fluctuations.

NOMINATION. Under a lifting agreement the amount of crude oil a working interest owner is expected to lift. Each working interest partner has a specific entitlement depending upon the level of production, royalties, their working interest, and their relative position (i.e., underlifted or overlifted), etc. Each working interest partner must notify the operator (nominate) the amount of its entitlement that it will lift. Sometimes, depending upon the lifting agreement, the nomination may be more or less than the actual entitlement. (see Liftings and Entitlements).

OWNER'S EQUITY. Also known as book value of the corporation. It is equal to assets minus liabilities.

OPERATING PROFIT (OR LOSS). The difference between business revenues and the associated costs and expenses (COGS and operating expenses) exclusive of interest or other financing expenses, and extraordinary items or ancillary activities. Synonymous with net operating profit (or loss), operating income (or loss), and net operating income (or loss).

OPIC (OVERSEAS PRIVATE INVESTMENT CORPORATION). A U.S. government agency founded under the Foreign Assistance Act of 1969 to administer the national investment guarantee program for investment in less developed countries (LDCs) through the issuance of insurance for risks associated with war, expropriation and inconvertibility of payments in local currency.

PARENT COMPANY. A company that owns controlling interest in another company.

PETRODOLLARS. Deposits of dollars by countries which receive dollar revenues due to the sale of petroleum to other countries; the term commonly refers to OPEC deposits of dollars in the Eurocurrency market.

PORTFOLIO. The aggregate of all assets held by an investor.

POSTED PRICE. The official government selling price of crude oil. Posted prices may or may not reflect actual market values or market prices.

PLATFORM SCENARIO. The desired endstate foreseen by strategic planning process and often based on managers' mental models of the most probable future.

PREFERRED STOCK. A type of stock that has preference over common stock usually in regard to payment of dividends.

PRESENT VALUE. The amount that, if paid today, would be equivalent to a future payment or stream of future payments based upon a specified interest (discount) rate. The sum of all discounted cash flows from a particular investment.

PRICE/EARNINGS RATIO OR P/E RATIO. The relationship of a stock price and the earnings per share defined as the price per share divided by the carnings per share.

PRIME LENDING RATE. Typically considered the interest rate on short-term loans banks charge to their most stable and credit-worthy customers. The prime rate charged by major lending institutions is closely watched and is considered a benchmark by which other loans are based. For example, a less well established company may borrow at prime plus 1%.

PRODUCTION SHARING AGREEMENT (PSA). This is the same as a Production Sharing Contract (PSC). While at one time this term was quite common, it is used less frequently now and the term Production Sharing Contract is becoming more common.

PROGRESSIVE TAXATION. Where tax rates increase as the basis to which the applied tax increases. Or, where tax rates decrease as the basis decreases. The opposite of regressive taxation.

PROPERTY, PLANT, AND EQUIPMENT. The tangible assets of a long-term nature used in continuing operations.

PRO FORMA. Latin for as a matter of form. A financial projection based upon assumptions and possible events that have not occurred. For example, a financial analyst may create a consolidated balance sheet of two nonrelated companies to see what the combination would look like if the companies had merged. Often a cash flow projection, for discounted cash flow analysis, is referred to as a pro forma cash flow.

PROVED RESERVES. Quantities of oil or natural gas that can with reasonable certainty be recovered economically with existing technology.

PUT OPTION. Grants the right to sell a particular asset at a specified price within a specified period of time.

QUICK RATIO. The more liquid current assets (cash, marketable securities, and accounts receivable) divided by current liabilities.

RACK PRICE. In the United States, gasoline has a rack price which is a wholesale price at a terminal or refinery truck loading facility, called a rack to nonbranded, independent marketers. The rack price is considered gasoline's commodity price.

REGIONAL DEVELOPMENT BANKS (RDB). Banks owned by governments of the corresponding regions and by governments of industrialized nations. They are designed to assist in reducing poverty and in promoting economic growth. Modeled after the World Bank with respect to procedures for project preparation and implementation.

RELINQUISHMENT. A contract clause that refers to how much of a contract or license area a contractor must surrender or give back to the government during or after the exploration phase of a contract. Licenses are usually granted on the basis of an initial term with specific provisions for the timing and amount of relinquishment prior to entering the next phase of the contract. Also referred to as exclusion of areas.

RESERVE REPLACEMENT RATIO. The amount of oil and gas discovered in a given period divided by the amount of production during that period.

RETAINED EARNINGS. Stockholder's equity that arises from accumulated earnings less dividends, losses, or transfers of contributed capital.

RETURN ON ASSETS. Measure of profitability; net income divided by total assets.

RETURN ON EQUITY. Measure of profitability; net income divided by shareholder equity.

REVENUE RECOGNITION. The process of determining when a sale takes place; a procedure of accrual accounting.

RINGFENCING. A cost center based fiscal device that forces contractors or concessionaires to restrict all cost recovery and or deductions associated with a given license (or sometimes a given field) to that particular cost center. The cost centers may be individual licenses or on a field-by-field basis. For example, exploration expenses in one nonproducing block could not be deducted against income for tax calculation in another block. Under a PSC, ringfencing acts in the same way—cost incurred in one ringfenced block cannot be recovered from another block outside the ringfence. The term *consolidaton* is lused in a similar context. Ringfencing means "no consolidation."

RISK PREMIUM. The return over and above the riskless return investors try to obtain as a compensation for the risk born by holding a particular investment.

ROYALTY HOLIDAY. A form of fiscal incentive to encourage investment and particularly marginal field development. A specified period of time, in years or months, during which royalties are not payable to the government. After the holiday period the standard royalty rates are applicable. (see Tax Holiday)

SCENARIO. A prediction of events leading to an endstate.

SCENARIO ENDSTATE. A snapshot of future industry conditions. In scenario planning these are designed to be purposefully extreme and divergent, capture bias and different points of view; designed to explore what drives what; and written from the point of view of the planning horizon year.

SCENARIO EVENTS. Events as defined in scenario planning can be influenced by industry participants. They are specific, concrete manifestations of trends and issues and can be the basis of an environmental monitoring and early warning system.

SCENARIO PLANNING. A tool for helping to take a long view in a world of great uncertainty. A prediction of events of the future which helps in recognizing and adapting to changing aspects of the present environment.

SECURED BONDS. Bonds that give the bondholders a pledge of certain assets of the company as collateral—guarantee of payment.

SEISMIC OPTION. A contractual arrangement or agreement between a host government and a contractor that gives the contractor a period during which it has exclusive rights over a geographic area or contract area during which the contractor will shoot additional seismic data. After data acquisition, processing, and interpretation, the contractor has the right to enter into an additional phase of the agreement or a more formal contract with the government for the area which usually includes a drilling commitment.

SENIOR SECURITIES. The classification of securities as senior indicates a first or primary claim in case of default. Secondary securities, sometimes called junior securities, claim follows those with primary or senior interest.

SEVERANCE TAX. A tax on the removal of minerals from the ground usually levied as a percentage of the gross value of the minerals removed. The tax can also be levied on the basis of so many cents per barrel or per million cubic feet of gas.

SINKING FUND. Money accumulated on a regular basis is a separate account for the purpose of paying off an obligation or debt.

SLIDING SCALES. A mechanism in a fiscal system that increases effective taxes, and/or royalties based upon profitability or some proxy for profitability such as increased levels of oil or gas production. Ordinarily, each tranche of production is subject to a specific rate and the term incremental sliding scale is sometimes used to further identify this.

SPOT MARKET. An open market for oil or gas where buyers and sellers bid or negotiate on prices in expectation of taking delivery of the product. This is different than the futures market where a contract is usually purchased with no intention of taking physical delivery.

SPOT PRICE. Oil or gas prices established in the open spot market. Often considered a leading indicator of price direction and contract prices.

STATE TAKE. The government share of profits also referred to as government take. There are a number of definitions, but the most succinct is: State Take equals State Cash Flow divided by Gross Project Cash Flow. This explicitly includes share of profits to the State from national oil company participation.

STOCK SPLIT. An increase in the number of outstanding shares of stock accompanied by a proportionate reduction in the par or stated value.

STRIPPER WELL. A well that produces at a low rate of production. For U.S. legal/tax purposes a stripper well is a well that produces on the average less than 10 barrels of oil per day or less than 60 thousand cubic feet of gas per day.

SUBSIDIARY. A company legally separated from but controlled by a parent company who owns more than 50% of the voting shares. A subsidiary is always by definition an affiliate company. Subsidiary companies are normally taxed as profits are distributed as opposed to branch profits which are taxed as they accrue. (see Affiliate)

SUCCESSFUL EFFORTS ACCOUNTING. A method of accounting for oil and gas exploration and development costs under which the costs of successful exploration are capitalized and the costs of unsuccessful exploration are expensed. (see Full Cost Accounting)

SUCCESS RATIO. Ratio of successful wells to total wells drilled. A distinction is sometimes made between technical success and commercial success for a well, where technical success simply refers to whether or not hydrocarbons have been found and commercial success refers to whether or not the hydrocarbons found were in commercial quantities or not.

SURRENDER. Surrender is most often used synonymously with relinquishment in the context of area reduction. However, the term also is used to describe a contractor's option to withdraw from a license or contract at or after various stages in a contract. (see Relinquishment)

SUNK COSTS. There are a number of categories of sunk costs:

- Tax Loss Carry Forward (TLCF)
- Unrecovered Depreciation Balance
- Unrecovered Amortization Balance
- Cost Recovery Carry Forward

These costs represent previously incurred costs that will ultimately flow through cost recovery or will be available as deductions against various taxes (if eligible).

SWAP. Agreement to exchange one asset for another at a specified rate and date. The banks commonly serve as intermediaries between two parties who wish to engage in a currency swap.

TAKE-OR-PAY CONTRACT. A type of contract where specific quantities of gas (usually daily or annual rates) must be paid for, even if delivery is not taken. The purchaser may have the right in following years to take gas that had been paid for but not taken.

TANGIBLE ASSETS. Long-term assets that have physical substance.

TANGIBLE COSTS. Costs associated with capital items which have a useable life and salvage value. These costs must be capitalized (or depreciated).

TAX. A compulsory payment pursuant to the authority of a foreign government. Fines, penalties, interest and customs duties are not taxes.

TAX CREDITS. Deductions from the computed tax liability.

TAX HOLIDAY. A form of fiscal incentive to encourage investment. A specified period of time, in years or months, during which income taxes are not payable to the government. After the holiday period the standard tax rates apply.

TAX TREATY. A treaty between two (bilateral) or more (multilateral) nations which lowers or abolishes withholding taxes on interest and dividends, or grants creditability of income taxes to avoid double taxation.

TRANSFER PRICING. Integrated oil companies must establish a price at which upstream segments of the company sell crude oil production to the downstream refining and marketing segments. This is done for the purpose of accounting and tax purposes. Where intrafirm (transfer) prices are different than established market prices, governments will force companies to use a marker price or a basket price for purposes of calculating cost oil and taxes. Transfer pricing also refers to pricing of goods in transactions between associated companies.

TRANSLATION ADJUSTMENTS. Changes in the financial statement due wholly to currency exchange rate fluctuations—referred to as translation gains or losses.

TREASURY STOCK. Capital stock of a company, either common or preferred, that was issued and reacquired by the company but has not been reissued or retired.

UNBUNDLING. A term which is used to describe the deregulatory effort by the Federal Energy Regulatory Commission (FERC) which began in 1983 and completed by FERC Order 636 issued in 1992. Prior to the unbundling process, pipelines were not only transporting but also purchasing gas from their suppliers and reselling to their customers, i.e., the merchant function. After the unbundling process, pipelines became more concerned with the transportation function for a fee and the producers or some other entity can now sell gas directly to the customer. Also, maximum lawful prices and regulations applicable to producer sales gradually came to an end.

UNIFORM COMMERCIAL CODE (UCC). In the interests of uniformity and reform, the National Conference of Commissioners on Uniform State Laws and the American Law Institute sponsored and directed preparation of the Uniform Commercial Code. The work on the UCC began in 1942, and the finished draft was completed in 1952. The UCC consists of ten articles:

1. General Provisions
2. Sales
3. Commercial Paper
4. Bank Deposits and Collections
5. Letters of Credit
6. Bulk Transfers
7. Documents of Title
8. Investment Securities
9. Secured Transactions
10. Effective Date and Repealer

UPLIFT. Common terminology for a fiscal incentive whereby the government allows the contractor to recover some additional percentage of tangible capital expenditure. For example if a contractor spent $10 million on eligible expenditures and the government allowed a 20% uplift, then the contractor would be able to recover $12 million. The uplift is similar to an investment credit. However, the term often implies that all costs are eligible where the investment credit applies to certain eligible costs. The term uplift is also used at times to refer to the built-in rate of return element in a rate of return contract.

UNIT-OF-PRODUCTION DEPRECIATION.

Formula for Unit-of-Production Method

$$\text{Annual Depreciation} = (C - AD - S)\ P \div R$$

where

C = Capital Costs of equipment

AD = Accumulated depreciation

S = Salvage value

P = Barrels of oil produced during the year *

R = Recoverable reserves remaining at the beginning of the tax year

* If there is both oil and gas production associated with the capital costs being depreciated, then the gas can be converted to oil on a thermal basis.

Unsecured Bonds. Bonds issued on the general credit of a company—debenture bonds.

Value-Added Tax. A tax that is levied at each stage of the production cycle or at the point of sale. Normally associated with consumer goods. The tax is assessed in proportion to the value added at any given stage.

Wasting Assets. A term used to apply to natural resources.

World Bank. Refers to the International Bank for Reconstruction and Development (IBRD) and it's affiliates: the International Development Association (IDA), International Finance Corporation (IFC), and Multilateral Investment Guarantee Agency (MIGA).

Working Capital or Net Working Capital. Current assets minus current liabilities. Working capital represents the minimum amount of cash a company could raise in a sudden liquidation.

Working Interest. The percentage interest ownership a company (or government) has in a joint venture, partnership, or consortium. The expense-bearing interests of various working interest owners during exploration, development and production operations, may change at certain stages of a contract or license. For example, a partner with a 20% working interest in a concession may be required to pay 30% of exploration costs, but only a 20% share of development costs. With government participation, the host government usually pays no exploration expenses, but pays pro-rata development and operating costs and expenses.

Zero Coupon Bonds ("Zeros"). Bonds with no periodic interest payment, just an obligation to pay a fixed amount (interest and principal) at the maturity date.

References

Barry, Richard. 1993. *The Management of International Oil Operations.* PennWell Publishing Co., Tulsa, Oklahoma.

Brock, Horace R, Jennings, Dennis R. and Feiten, Joseph B. 1996. *Petroleum Accounting, Principles, Procedures, & Issues,* 4th ed. Professional Development Institute. Denton, Texas.

Fogg, C. Davis. 1994. *Team-Based Strategic Planning.* American Management Association, New York.

Gleim, Irvin N. and Flesher, Dale L. 1996. *Certified Financial Management Review.* Gleim Publications, Gainesville, Florida.

Johnston, Daniel. 1992. *Oil Company Financial Analysis in Nontechnical Language*, PennWell Books, Tulsa, Oklahoma.

Johnston, Daniel. 1994. *International Petroleum Fiscal Systems and Production Sharing Contracts*, PennWell Books, Tulsa, Oklahoma.

Kaplan, Robert S., and Norton, David P. 1996. *The Balanced Scorecard.* Harvard Business School Press, Boston.

Kieso, Donald E. and Weygandt, Jerry J. 1995. *Intermediate Accounting,* 8th ed. John Wiley & Sons, Inc., New York.

Loscala, William. 1982. *Cash Flow Forecasting.* Prentice Hall Business & Professional Division, Englewood Cliffs, New Jersey.

Madura, Jeff. 1992. *International Financial Management*, 3rd ed. West Publishing Company, New York.

Megill, Robert E. 1988. *An Introduction to Exploration Economics,* 3rd ed. PennWell Books, Tulsa, Oklahoma.

Monroe, Ann. 1994. *Bridge Over Troubled Waters*. CFO Publishing Co. Boston.

Morgan, Franger M. July 1993. *"Risk Analysis and Management."* Scientific American Magazine.

Newendorp, Paul. 1975. *Decision Analysis for Petroleum Exploration.* PennWell Publishing Co., Tulsa, Oklahoma.

Pegado, Alexandre Do Rego Pinto. 1989. *World Crude Oil Trading Agreements & Procedures.* Barrows Company, New York.

Ross, Stephen A., Westerfield, Randolph W., and Jaffe, Jeffrey F. *Corporate Finance,* 3rd ed. Richard D. Irwin, Inc., Homewood, Illinois, 1993.

Schuyler, J.R. 1996. *Decision Analysis in Projects.* Project Management Institute, Upper Darby, Pennsylvania, pg. 144.

Siegel, Dr. Joel G. 1982. *How to Analyze Businesses, Financial Statements and the Quality of Earnings.* Prentice Hall Inc., Englewood Cliffs, New Jersey.

Steinmetz, R., ed., Megill, Robert E. 1992. "Estimating Prospect Sizes." "The Business Side of Petroleum Exploration." "Treatise of Petroleum Geology." *Handbook of Petroleum Geology*. AAPG, pg. 63–69.

Stewart, G. Bennett. 1991. *The Quest for Value.* Harper Collins, Publishing, Inc. New York.

Thompson, Robert S. and Wright, John D. 1985. *Oil Property Evaluation,* 2nd ed. Thompson-Wright Associates.

Treat, John Elting. 1995. *Creating the High Performance International Petroleum Company: Dinosaurs Can Fly.* PennWell Publishing Co., Tulsa, Oklahoma.

Weston, J. Fred and Copeland, Thomas E. 1992. *Management Finance,* 9th ed. The Dryden Press, Chicago.

Willson, James D, Roehl-Anderson, Janice, Bragg, Steven M. 1995. *Controllership: The Work of the Managerial Accountant,* 5th ed. New York.

Index

D

E

F

G

H

I

J

L

M

N

O

P

T

U

V

W

Z